SQL Data Analytics Made Easy: Your Step-by-Step Guide to Unlocking Data's Hidden Secrets

Dive into the world of data analytics with SQL, demystify complex concepts, and harness the power of data to drive intelligent decision-making effortlessly

L.D. Knowings

Contents

Introduction

"Imagine a world overflowing with data – numbers, stats, facts, all buzzing around, waiting to tell a story." This line captures the vivid reality of our modern, data-driven world. It's a world that may seem overwhelming at first, but there's a way to make sense of it all and unravel the tangled mesh of information. SQL, or Structured Query Language, is the magical compass that can guide us to reveal the hidden treasures within data. It's not merely a tool for tech geeks or data scientists; it's a language that anyone, including you, can learn and use to decode the secrets hidden in numbers.

SQL's Staying Power

SQL has stood the test of time. Though it's been around for several decades, it's still among the most common programming languages, consistently ranking among the top 10 most-used programming languages. You can check online lists such as **Statistics Times** or **StackScale** to see its enduring presence.

Why does SQL continue to thrive in an ever-changing technological landscape? The answer lies in its universal applicability and robust functionality. Unlike some programming languages that fade into obscurity, SQL's relevance grows with the ever-increasing emphasis on data analytics.

A Language for All

Let's pause momentarily and reflect: What do you think of when you hear the word "programming language"? For many, it might evoke images of complex coding, mathematical equations, and an exclusive domain for tech experts. But SQL is different. It's a language that speaks to data analysts, business professionals, researchers, and even curious individuals fascinated by the power of data.

Do you recall when you had to sift through a large spreadsheet, looking for specific information? Imagine if that spreadsheet were the size of a city, filled with rows and columns stretching as far as the eye can see. Finding what you need would be daunting without the right tools.

That's where SQL comes in. The tool turns an overwhelming city of data into a navigable map, allowing you to find exactly what you're looking for precisely and efficiently.

Breaking Down Barriers

One of the common misconceptions about SQL is that it's too technical, math-heavy, and requires advanced programming skills. It's a misconception that has deterred many from exploring this powerful language. But here's the truth: SQL is accessible.

This book is designed to break down those barriers, taking you from the basics to hands-on problem-solving without overwhelming you with jargon or complex mathematics. It's a guide that doesn't just teach you the "how" and the "why" behind each concept, facilitating a deeper grasp of SQL and data analytics.

Real-World Applications

SQL is not confined to abstract exercises or theoretical knowledge. It's a language used across various industries such as technology, finance, marketing, healthcare, etc. Whether you're a data analyst seeking to enhance your professional skills or a business owner looking to make data-driven decisions, SQL is a tool that can be applied in real-world scenarios.

Consider the case studies and practical examples included in this book. They highlight how SQL can effectively solve real-world problems, from integrating data from multiple sources to visualizing complex datasets for actionable insights.

Introduction

The Challenges in the World of Data

"Data is the new oil." You've probably heard this phrase more times than you can count, and while it might seem like a buzzword, the analogy holds some truth. Just like oil, data must be refined and processed to become valuable. And this refining process is where many of us stumble.

Struggling to Make Sense of Data

If you've ever found yourself staring at a spreadsheet filled with numbers, symbols, and text, feeling lost and overwhelmed, you're not alone. The amount of data available today is staggering, and the task of sifting through it to find meaningful insights can be daunting.

I remember working on a project that required me to analyze customer behavior data. At first glance, the dataset was chaotic, with thousands of rows and columns. The challenge was not just understanding the data but transforming it into actionable insights. It was a puzzle that required patience, skill, and a keen eye for patterns.

Difficulty in Extracting Meaningful Insights

But it's not just about understanding the data; it's about extracting insights that can drive intelligent decision-making. You may have all the data, but it remains just a pile of numbers if you can't turn it into something meaningful.

Introduction

Have you ever had that moment where you've gathered all the information and crunched the numbers, but the 'Eureka' moment seems elusive? It's a common challenge, making data analysis feel frustrating and futile.

Overwhelmed by Complexity

And then data analysis tools are complex. SQL is a powerful language but can also be intimidating for newcomers. Add other software, platforms, and methodologies, and you can quickly feel overwhelmed.

Maybe you've tried to teach yourself SQL through online tutorials, only to get lost in the technical jargon. Perhaps you've attempted to integrate data from multiple sources, only to end up with inconsistencies and errors. These experiences can be disheartening and may even lead you to question whether you're cut out for the world of data analytics.

You're Not Alone

If these challenges resonate with you, know you're not alone. These struggles are shared by many, from seasoned professionals to curious beginners. The world of data is complex and ever-changing, but it's also exciting and full of potential.

Introduction

The Catalyst: Why This Book?

So why this book? What brought you here? The answer to that question is as unique as you are. But let's explore some common catalysts that might have triggered your interest in this book and the world of SQL and data analytics.

Enhancing Your Career

Maybe you're a professional looking to take your career to the next level. In an increasingly data-driven world, proficiency in data analytics is no longer a 'nice-to-have' but a 'must-have.' Whether you're a data analyst, business professional, or researcher, mastering SQL can open doors to new opportunities and set you apart from the competition.

Competing in the Job Market

Or perhaps you're seeking to make yourself more competitive in the job market. The demand for data-savvy individuals is on the rise, and having SQL and data analytics skills on your resume can give you an edge. It's not just about landing a job; it's about positioning yourself as a valuable asset in a rapidly evolving industry.

Effectiveness in Your Current Role

Maybe you're already in a role where data plays a crucial part and want to become more effective and efficient.

Introduction

Understanding SQL and data analytics can transform how you approach problems, make decisions, and communicate insights. It's about leveraging data to drive success in business, healthcare, marketing, or any field that relies on informed decision-making.

Satisfying Your Curiosity

Or maybe you're here out of sheer curiosity. The world of data is fascinating, and the ability to decipher it to uncover hidden patterns and insights is a skill that can be applied in various aspects of life. Whether you're curious about how companies target advertisements or how healthcare providers use data to improve patient care, understanding SQL and data analytics can enrich your understanding of the world around you.

Connecting with You

No matter what brought you here, know I understand where you're coming from. I've faced the same challenges, asked the same questions, and felt the same curiosity you might feel right now.

This book is not just a guide to SQL; it's a companion on your path to unlocking the hidden secrets within data. It's about breaking down barriers, demystifying complex concepts, and guiding you to harness the power of data to drive intelligent decision-making effortlessly.

Introduction

The Benefits of Learning SQL Through This Book

> *"The future belongs to those who understand that doing more with less is compassionate, prosperous, enduring, and thus more intelligent, even competitive."*

> — Paul Hawken

Practical Skill Acquisition

The world of data is vast, and understanding how to navigate it can be the difference between chaos and clarity. SQL, a language designed to communicate with databases, is your compass in this world. This book will teach you practical skills, not just theoretical knowledge.

Remember those days when data seemed like an impenetrable fortress? Through hands-on exercises, real-world case studies, and step-by-step guidance, you'll learn how to unlock the doors to data's hidden treasures.

Understanding of Data Analysis and Manipulation

Data is more than just numbers and text. It tells stories, reveals patterns, and can guide decision-making. With SQL, you'll learn to analyze and manipulate data, transforming raw information into meaningful insights.

Introduction

Imagine looking at a complex dataset and seeing beyond the surface. You'll uncover trends, identify anomalies, and better understand what the data tells you. It's like conversing with data, and SQL is the language you'll use to communicate.

Confidence in Handling Complex Datasets

Complex datasets can be intimidating, but mastering SQL gives you the confidence to handle them easily. No longer will you be overwhelmed by rows and columns; instead, you'll see opportunities to explore, analyze, and interpret.

Think back to when you felt lost in a sea of data. With SQL, you'll have the tools to navigate that sea, to find the hidden gems, and to make sense of the seemingly incomprehensible. Your relationship with data will transform from one of apprehension to one of mastery.

Ability to Make Data-Driven Decisions

In today's world, data-driven decisions are the cornerstone of success. Whether in business, healthcare, finance, or technology, making informed choices based on solid data analysis is crucial.

With SQL, you'll learn how to access and analyze data and use it to make intelligent decisions. Imagine being able to support your choices with concrete evidence, to back your strategies with data, and to lead with confidence. That's the power of SQL, which you'll gain through this book.

Potential Career Advancement

The demand for data-savvy professionals is rising, and SQL is a skill that can set you apart. Mastering SQL can open doors if you want to advance in your current role or explore new opportunities.

Consider the possibilities that await you with this newfound expertise. You'll be more valuable to your organization, more competitive in the job market, and more empowered to pursue your career goals. It's not just about learning a programming language; it's about investing in your future.

Painting a Picture: The Transformation

Your relationship with data is about to change. No longer will you see it as a chaotic mess, a puzzle with missing pieces, or a challenge too great to overcome. With SQL, you'll see data as a source of valuable insights, a pathway to understanding, and a tool for success.

You'll move from confusion to clarity, from uncertainty to confidence, from being a passive observer to an active participant in the world of data. It's not just about learning a skill; it's about embracing a mindset that empowers you to see the potential in data, unlock its secrets, and harness its power.

Imagine yourself in a room filled with scattered papers, numbers, graphs, and charts. Before, this room might have seemed overwhelming, where you could quickly lose yourself. But now, you see patterns, connections, and insights with SQL. The space transforms into a treasure trove of information for you to explore.

Your newfound skills will help you navigate this room and equip you to build bridges, connect dots, and create narratives that make sense of the data. You'll become a storyteller, a problem-solver, a decision-maker, and a leader.

A Brief Snapshot of What This Book Covers

 "Data is a precious thing and will last longer than the systems themselves."

— Tim Berners-Lee

From the Basics of SQL to Advanced Queries

There's a magic in understanding data, a power in knowing how to communicate with it. It's the language of the future, and SQL is one of its essential dialects. If you've ever felt overwhelmed by databases or mystified by queries, this book is your guide.

Introduction

Starting with the fundamentals, we'll explore the basics of SQL together. You'll learn how to write simple queries, retrieve data, and make sense of the numbers and text you encounter. But we won't stop there. Together, we'll move into more advanced territory, exploring complex queries and diving into the intricacies of data manipulation. It's rich and full of potential worlds, and I'm here to guide you every step of the way.

Data Integration and Statistical Analysis

Data doesn't always come neatly packaged. Sometimes, it's scattered across different sources, and bringing it together can be daunting. In this book, you'll discover the art of data integration, combining and unifying data into a coherent whole. You'll also learn statistical analysis techniques, transforming raw data into meaningful insights.

Remember those statistics classes that seemed so dry and theoretical? Here, we'll apply those concepts practically, making sense of numbers, finding patterns, and uncovering the hidden stories in data.

Data Visualization and Real-world Applications

Visualizing data is like painting a picture with numbers. It's about telling a story, about making complex information accessible and engaging. In this book, you'll learn the craft of data visualization, turning rows and columns into graphs, charts, and visual narratives.

But what's the use of a skill if it's not applied in the real world? Here, you'll find real-world applications and case studies demonstrating how SQL is used across various industries. SQL is a tool that transcends boundaries from healthcare to finance, marketing to technology. It's a skill that's applicable, relevant, and valuable.

Tips for a Successful Career in SQL

Your aspirations, your career goals, your desire to grow and succeed—all of these are part of this book. Whether you want to advance in your current role or explore new opportunities, mastering SQL can be a game-changer.

I'll share tips, insights, and guidance on how to build a successful career in SQL. It's not just about learning a programming language; it's about understanding the landscape, knowing the trends, and positioning yourself for success.

This Is the Right Book for You

You might be wondering if this book is for you. Maybe you're new to SQL or have some experience but want to deepen your understanding. Perhaps you're a professional looking to enhance your skills or a curious individual eager

Introduction

to explore the world of data. Whoever you are, whatever your background, this book is your companion.

I know you have ambitions. I know you have curiosity. I know you want to unlock the hidden secrets of data, transform chaos into clarity, and turn questions into answers.

I know this because I've been where you are. I've felt the excitement of discovery, the thrill of mastery, the joy of growth. And I want to share that journey with you.

So here it is an invitation, a promise, a commitment. This book is not just a collection of pages and words; it's a pathway to understanding, a tool for success, and a resource tailored for you.

Let's explore together. Let's learn, grow, and succeed. Let's make sense of data and make it sing and ours. This book is the perfect resource for you, and I'm here, ready to guide you. The world of SQL awaits, and it's yours for the taking.

Chapter 1
Introduction to Data Analytics and SQL

"Data is the new oil." You've probably heard this phrase, and it captures the essence of our modern world. But unlike oil, data is not finite; it's expanding at an unimaginable pace, becoming the lifeblood of businesses and industries across the globe. And the tool to refine this vast reserve? SQL (Structured Query Language) has become synonymous with data manipulation and analysis. This chapter is not just an introduction to data analytics and SQL; it's an invitation to a world that thrives on insights, decisions, and innovations, all fueled by data.

The Ever-Present Data

How many times have you checked your phone today? How many online purchases, social media posts, or GPS-guided routes have you made? Each action, each click, is a data point. According to a report by IDC, the world's data will grow to 175 zettabytes by 2025, a number so vast it defies comprehension. But it's not the sheer volume that's impressive; it's what this data can do.

Imagine a hospital integrating patient records, lab results, and treatment plans to provide personalized care. Think of a retailer predicting your next purchase based on browsing history and shopping behavior. These scenarios are not futuristic dreams but realities made possible through data analytics.

SQL: The Lingua Franca of Data

If data is the oil, then SQL is the refinery. It's the language that helps transform raw data into meaningful insights. Whether you're a data scientist trying to predict the next big trend or a business analyst aiming to understand customer behavior, SQL is your go-to tool.

Why has SQL become so indispensable? It's because of its simplicity and universality. From small startups to Fortune 500 companies, SQL is a common language that enables professionals to interact with databases, regardless of their complexity or size.

The Many Faces of Data Analytics

Data analytics is not a monolithic field; it's a rich tapestry of techniques and applications. It's about more than just crunching numbers; it's about telling stories, solving problems, and driving innovation.

- Descriptive Analytics: Understanding the past by summarizing data into meaningful insights. What happened last quarter with sales? Why did the website traffic spike the previous weekend?
- Predictive Analytics: Anticipating the future based on historical data. Can we forecast the next big product? What's the likelihood of a customer churning?
- Prescriptive Analytics: Recommending actions based on data-driven insights. Should we increase marketing spend in a particular region? How do we allocate resources for maximum efficiency?

These layers of analytics are what make the field so dynamic and exciting. They provide a pathway for individuals and organizations to move from merely collecting data to actively using it to make informed decisions.

Hands-On with SQL

The beauty of SQL lies in its accessibility. You don't have to be a computer scientist or a mathematician to master it. With a structured approach and hands-on practice, anyone can learn to write queries, analyze data, and uncover hidden patterns.

Throughout this book, we'll explore SQL in depth, starting with basic queries and gradually progressing to more complex operations. We'll also dive into real-world applications, allowing you to see SQL across various industries.

Connecting SQL with Your Career

Perhaps you're a business analyst seeking to enhance your reporting capabilities. Maybe you're a researcher striving to make sense of vast datasets. Or perhaps you're a curious individual eager to explore the world of data analytics. Regardless of your background or goals, SQL is a skill that can propel your career forward.

The demand for professionals with SQL expertise is growing, and it transcends traditional tech roles. From healthcare to marketing, finance to education, the ability to analyze data with SQL is becoming a valuable asset.

What is Data Analytics?

"Without data, you're just another person with an opinion." These words from W. Edwards Deming resonate profoundly in today's data-driven world. Let's take a moment to reflect on a seemingly simple decision: choosing a movie on a streaming service. You browse through categories, maybe watch a trailer or two, and select a film that catches your eye. But did you know that every click, every pause, every selection feeds into a complex system that analyzes your preferences, compares them with others, and recommends movies tailored to your taste? Welcome to the world of data analytics - a universe where raw data transforms into actionable insights.

A Simple Definition for a Complex Process

Data analytics is more than a buzzword; it examines, cleans, and transforms raw data to discover useful information. In layman's terms, imagine sifting through a pile of puzzle pieces, discarding the irrelevant ones, and carefully arranging the remaining pieces to reveal a complete picture. That picture could be your customer's buying habits, a patient's health trend, or a city's traffic pattern. The essence of data analytics lies in turning chaos into clarity, obscurity, and understanding.

The Benefits That Reach Beyond Business

The impact of data analytics isn't confined to boardrooms and balance sheets. Its tendrils reach every aspect of our lives, shaping decisions and strategies across various fields.

- In Business: Companies use data analytics to understand their customers better, streamline operations, and make informed decisions. A study by MIT showed that businesses that use data-driven decision-making are 5% more productive and 6% more profitable than their competitors.
- In Healthcare: Clinicians and healthcare providers analyze patient data to personalize treatments,

improve care quality, and predict outbreaks. Analytics in healthcare is expected to save $300 billion annually, particularly in the area of chronic disease prevention.

- In Sports: Coaches and athletes leverage data analytics to optimize performance, strategize plays, and scout talent. The use of analytics in sports has grown by 200% in the last three years, revolutionizing the game both on and off the field.

The list encompasses education, entertainment, law enforcement, and more. The common thread is the ability to make decisions based on facts, patterns, and insights rather than intuition or guesswork.

Stages of Data Analytics: From Chaos to Clarity

The journey from raw data to meaningful insight is not a straight path but a meticulously planned process. Here's a glance at the stages that turn data into decisions:

1. Data Collection: Gathering information from various sources, such as social media, sensors, or customer interactions. For example, social media platforms collect user behavior, likes, shares, and more data.
2. Data Processing: Cleaning and organizing the collected data to filter out noise and irrelevant

information. This step ensures that only valuable data is analyzed.

3. Data Analysis: Applying statistical methods and tools to spot patterns and relationships within the data. In our social media example, this could mean identifying trends in user engagement or content popularity.
4. Data Interpretation: Translating the statistical findings into actionable insights. This stage makes the analysis human-readable, connecting the dots to form a cohesive narrative.
5. Decision Making: Using the interpreted data to make informed decisions, such as targeting ads based on user preferences or adjusting content delivery.

Each stage plays a vital role, and missing or skimping on one can lead to misguided insights or lost opportunities.

Real-World Examples: A Closer Look

Let's delve into a real-world scenario where data analytics shines. Consider a leading e-commerce platform. They collect data from millions of users on what they click, buy, and leave in their carts. By processing this data, they filter out irrelevant information, focusing only on user purchasing behavior. Analyzing this behavior, they spot patterns: preferences for specific brands, seasonal buying trends, and responsiveness to discounts. Interpreting these patterns, they craft personalized recommendations, flash sales, or loyalty programs. The result? Increased sales, satisfied customers, and a thriving business.

The Crucial Role of SQL in Data Analytics

"The future is made of data, and the future is now." These words encapsulate the essence of data analytics, and at the core of this pulsating future lies a language, a tool that turns data into insights: SQL or Structured Query Language. If data is the raw material, SQL is the master craftsman, meticulously carving wisdom out of chaos. Let me guide you through SQL and why it is a beacon in data analytics.

A Language That Speaks to Data

SQL, or Structured Query Language, is like the grammar of data. It's a standardized language specifically designed to communicate with databases. Think of it as the bridge between human curiosity and machine intelligence, a way to ask a database questions and get precise answers. In the vast universe of data, SQL is the translator, the interpreter that turns human inquiries into commands that databases can understand.

The Legacy of SQL: A Brief Glimpse into History

The genesis of SQL dates back to the 1970s when IBM engineers Raymond Boyce and Donald Chamberlin were grappling with the challenge of managing and manipulating complex datasets. They envisioned a language that could converse with data, not in bits and bytes but in simple, intuitive commands. Thus, SQL was born.

SQL has evolved, adapted, and expanded since those early days, becoming the lingua franca of data management. It's no longer confined to the hallways of tech giants but is a part of every industry and every organization that deals with data. In the 50 years since its inception, SQL has not just survived but thrived, cementing its position as an indispensable tool in the modern data landscape.

SQL: The Powerhouse of Data Querying and Manipulation

Why does SQL hold such a revered place in data analytics? SQL does more than just store data; it makes data meaningful. Allow me to explain:

1. Querying Data: SQL lets you ask complex or straightforward questions and get answers from a database. Want to know the sales trends of the last quarter? Need to identify the most common customer complaints? SQL lets you do that with a few lines of code.

2. Manipulating Data: Beyond just reading data, SQL allows you to add, update, delete, or modify information within a database. It's like having the

keys to a vast library, where you can read the books and add new ones, rearrange shelves, or even create new sections.

3. Connecting Data: SQL helps you link different pieces of data, uncovering relationships and patterns that may not be apparent at first glance. It's about connecting the dots and finding the hidden threads that weave the fabric of insights.

4. Securing Data: In an age where data privacy is paramount, SQL ensures that only authorized users can access or modify specific parts of a database. It's not just about retrieving information; it's about doing it responsibly and securely.

The Unrivaled Benefits of SQL

SQL's strength lies not just in its functionality but in its virtues that make it a preferred choice for data professionals:

- Wide Acceptance: SQL is almost universal. It's supported by significant database systems, making it a versatile choice.
- Easy Readability: Unlike many programming languages, SQL's commands are often simple and intuitive, resembling natural language.
- Strong Performance: SQL's performance is commendable, especially with large databases, allowing for efficient querying and manipulation.

- Scalability and Flexibility: SQL adapts and scales according to the needs of small businesses to multinational corporations, never losing efficiency.

SQL Across Industries

"Data is the new oil, and SQL is the machinery that refines it." This powerful statement encapsulates how SQL is not just a tool but a transformative force that permeates every facet of the modern world. Let's explore together the extraordinary reach of SQL across various industries and job roles and how this seemingly technical language has become a universal key to success.

SQL: The Universal Language of Industries

SQL, short for Structured Query Language, has transcended its technological origin to become integral to numerous sectors. Let's delve into the fascinating world of SQL's applications:

1. Technology Companies: Tech giants like Google, Amazon, and Facebook depend on SQL to manage vast troves of user data. From personalizing user experiences to optimizing advertising, SQL is at the heart of their operations.
2. Healthcare Facilities: Hospitals and healthcare organizations use SQL to handle patient records, track medical histories, and coordinate treatments.

It ensures that life-saving information is available at a moment's notice.

3. Financial Institutions: Banks and investment firms leverage SQL from risk management to customer service. It aids in analyzing market trends, managing portfolios, and ensuring regulatory compliance.

4. Retail and E-commerce: SQL helps retailers analyze consumer behavior, manage inventories, and forecast sales. It transforms raw sales data into actionable insights that drive business strategies.

5. Manufacturing Sector: In manufacturing, SQL is employed to monitor production lines, manage supply chains, and optimize resource allocation. It's the invisible hand that keeps the wheels of industry turning smoothly.

6. Educational Institutions: Universities and schools use SQL to manage student records, track academic progress, and facilitate online learning. It's the backbone of modern education's digital transformation.

7. Government Agencies: From law enforcement to social services, government bodies rely on SQL to manage public records, coordinate services, and analyze policy outcomes. It's a tool for governance in the information age.

Chapter 1

SQL in Various Job Roles: Not Just for Data Geeks

SQL is not confined to the cubicles of data analysts or database administrators. Its applications are so pervasive that professionals across various roles find themselves using SQL:

- Financial Analysts use SQL to query financial databases, assess market trends, and make investment recommendations.
- Project Managers: SQL aids in tracking project progress, managing resources, and generating progress reports.
- Marketing Professionals: SQL is a marketer's ally in understanding consumer behavior, from segmenting customers to tracking campaign performance.
- Healthcare Administrators: Managing patient records, scheduling appointments, and coordinating care, SQL is at the heart of healthcare administration.
- Logistics Managers: SQL helps track shipments, manage inventories, and optimize supply chain operations.

The breadth of SQL's applications underscores that it's not merely a technical skill; it's a professional asset, a tool that adds value to almost any job role.

SQL: Your Competitive Edge in the Job Market

In a world driven by data, proficiency in SQL is not just an advantage; it's a necessity. Here's why:

- High Demand: The demand for SQL skills transcends industries and job roles. A recent survey shows SQL ranks among the top ten most in-demand skills across job listings.
- Versatility: Whether in tech, finance, healthcare, or any other field, SQL is likely a part of your industry's toolset. It's a skill that opens doors and breaks down barriers.
- Career Growth: SQL proficiency can accelerate your career progression, opening opportunities for specialized roles like data scientists, business intelligence analysts, and more.
- Future-Proofing: SQL's relevance will only increase as data continues to shape the future. It's a skill that aligns with the future, ensuring you stay ahead of the curve.

SQL in the World of Data Analytics

"Data tells a story. SQL is the language that narrates it." With this insight, I invite you to join me in exploring the symbiotic relationship between SQL and the broader world of data analytics. It's more than just a connection between a programming language and a field of study; it's a collaboration that has shaped how we understand and interact with the world around us.

The Essential Role of SQL in the Data Analysis Pipeline

The data analysis pipeline is complex and intricate, encompassing everything from data collection to visualization. It's like assembling a puzzle; each piece must fit perfectly to reveal the complete picture. And SQL? SQL is the tool that helps you place those pieces in the proper order.

1. Data Collection and Preparation: Think of SQL as the sturdy foundation of your data analysis building. You use it to gather, clean, and organize your data, preparing it for analysis. Whether working with structured data from a database or pulling in data from various sources, SQL's robust querying capabilities ensure that your data is precise and relevant.

2. Data Analysis and Exploration: Now, imagine SQL as your guiding compass, helping you navigate the complex maze of data. It allows you to sift through millions of rows, filter out the noise, and zoom in on the insights that matter. SQL equips you with the tools to dissect and understand your data, from essential statistical summaries to complex aggregations.

3. Data Integration and Transformation: SQL plays the role of a master chef, blending different ingredients (data) into a cohesive dish (dataset). It enables you to join tables, merge datasets, and create new variables, ensuring your data is unified and ready for modeling.
4. Data Visualization: SQL is the artist's brush, turning raw numbers into vivid visualizations. Though often associated with tables and queries, SQL can also create charts, graphs, and dashboards, translating data into visuals that tell a compelling story.

SQL's Harmonious Interface with Other Data Analysis Tools

SQL doesn't work in isolation. It's part of a vibrant ecosystem of tools and technologies that form the data analytics landscape together. And it's in this collaborative environment that SQL truly shines.

- SQL and Python: Python is a popular programming language in data science, known for its versatility and ease of use. SQL complements Python by allowing you to query databases and integrate the results directly into your Python scripts. It's like conversing between two brilliant minds, each amplifying the other's strengths.

- SQL and Machine Learning: Machine learning models thrive on data, and SQL ensures that the data is tailored to the model's needs. Whether it's preprocessing the data, selecting the right features, or fetching real-time data for predictions, SQL is the bridge that connects machine learning algorithms to the data they require.
- SQL and Big Data Technologies: SQL continues to be relevant in the era of big data. Tools like Apache Hive and Spark SQL bring the power of SQL to big data processing, allowing you to query and analyze vast datasets with the familiar syntax of SQL.

Future Trends: SQL's Unwavering Relevance in a Data-Driven World

Data analytics is ever-evolving, shaped by emerging trends and technological advancements. But amid this constant flux, SQL stands as a timeless pillar, unwavering in its relevance and significance.

1. Adoption of SQL in Non-Traditional Domains: SQL is breaking free from its traditional confines, finding applications in marketing, HR, and sports analytics. It's becoming a universal language, transcending technical barriers and reaching professionals across various fields.
2. Integration with Modern Data Analysis Tools: The future will see closer integration between SQL and

modern data analytics tools, from machine learning libraries to cloud-based analytics platforms. SQL will continue to be the glue that binds these diverse technologies together.

3. Focus on Real-Time Analytics: As businesses seek to harness real-time insights, SQL adapts to real-time data processing and analytics needs. Tools like Apache Kafka and streaming SQL pave the way for SQL's role in real-time decision-making.

Segue: A New Chapter Awaits

We've explored the intricate relationship between SQL and data analytics, delved into its role in the data analysis pipeline, and reflected on its enduring presence in a dynamic data landscape. But this is just the beginning.

Ahead lies a new chapter, a new adventure, and a new opportunity. It's a chapter filled with discovery, creativity, and empowerment. It's where you'll take your first steps into the world of SQL, learn to create and manipulate databases, understand SQL syntax, and write your very first SQL queries.

You'll be more than just a learner; you'll be a creator, an explorer, and a storyteller. You'll have the power to transform data into insight, information into wisdom, and curiosity into innovation.

Chapter 1

Get ready to embrace the world of SQL, where endless possibilities await. Let's take this step together with confidence and enthusiasm. Chapter 2, "Getting Started with SQL," is on the horizon, calling your name. Are you ready to answer the call?

Chapter 2
Getting Started with SQL

"Why do we need a language to speak to our data?" It's a question that may seem peculiar at first glance, but it unlocks the essence of structured query languages. As you and I communicate through words and expressions, our data also needs a medium to express itself, reveal its secrets, and answer our questions. That medium is SQL. In this fascinating dialogue, let's discover how to converse with data and how SQL can become your ally.

The ABCs of SQL: Conversing with Data

SQL, or Structured Query Language, is more than a set of commands and syntax; it's a conversation with data. It's about asking questions, seeking answers, and finding meaning. When you write an SQL query, you're not just typing code; you're expressing a thought, a curiosity, a desire to know.

Databases and Tables: Consider databases as vast libraries and tables as books filled with books. Each book contains a story, which is a piece of data. SQL helps you navigate these shelves, pick the right books, and read the stories you're interested in.

Key Commands: There are specific commands like SELECT, INSERT, UPDATE, and DELETE that allow you to interact with these stories. You can read them (SELECT), add new ones (INSERT), modify them (UPDATE), or even remove them (DELETE). It's a language designed to cater to your curiosity.

Your First SQL Queries: A Gentle Introduction

It's time to take your first steps into the world of SQL. Don't worry; I'm here with you, guiding you through each step. Remember, it's not about memorizing commands but understanding the conversation you want with the data.

Creating a Database: Imagine you're building a new library and want to name it. In SQL, you'll use a simple command like CREATE DATABASE LibraryName; to give life to your library.

Creating Tables: Next, you'll develop shelves or tables to organize your books or data. A command like CREATE TABLE Authors (Name VARCHAR(50), Age INT); will create a shelf for authors with their names and ages.

Inserting Data: It's time to fill those shelves with books or data. You can add an author using INSERT INTO Authors (Name, Age) VALUES ('L.D. Knowings', 53);.

Querying Data: Finally, you'll learn to ask questions, like finding all authors above a certain age. SELECT * FROM Authors WHERE Age > 40; will provide you with the answers you seek.

Exercise: Your Connection with SQL

Please reflect on your connection with SQL. Grab a pen and paper and think about the following:

1. What drew you to SQL? Was it a personal project, a professional need, or sheer curiosity?
2. What questions do you want to ask your data? Think of specific queries that intrigue you.
3. How can SQL help you achieve your goals? Reflect on how mastering SQL can enhance your skills and career.

The Importance of SQL in Modern Data-Driven Decisions

We live in a world where data influences every aspect of our lives. From business decisions to medical research, data is at the core. SQL is the bridge that connects decision-makers to this invaluable resource.

Unlocking Insights: SQL allows you to dig deep into data, uncover patterns, and derive insights that can drive intelligent decisions. It's not just about numbers; it's about understanding the story behind them.

Empowering Individuals: SQL democratizes data. It makes it accessible to everyone, not just the tech-savvy. Whether you're a marketer, a researcher, or a small business owner, SQL empowers you to leverage data in your domain.

Fostering Collaboration: SQL encourages collaboration between different teams and disciplines. It's a universal language that breaks down silos and enables cross-functional synergy.

Databases and Tables

"A place for everything, and everything in its place." This age-old adage applies to your home or office organization and forms the foundation for managing, organizing, and accessing data in SQL. When you enter the realm of databases and tables, you embrace a system that ensures efficiency, accuracy, and simplicity. Let's explore this foundational concept by drawing parallels with everyday scenarios.

A World Organized: Understanding Databases and Tables

Imagine a sprawling library filled with millions of books. Finding a specific book would be like searching for a needle in a haystack without a systematic organization. Similarly, in the digital world, databases serve as the library, and tables act as shelves, organizing data in a way that's both logical and accessible.

Databases: A database is a collection of related data organized for convenient access. Think of it as the entire library, housing different sections or tables.

Tables: Tables are like shelves within the library, each dedicated to a specific subject. They consist of rows and columns, where rows represent individual records, and columns represent fields or attributes.

For example, a table named 'Students' might have columns like 'StudentID,' 'Name,' 'Age,' and 'Course,' while the rows would contain the actual details of each student.

The Building Blocks: Key Terminology

Understanding the language of databases and tables is essential. Here's a breakdown of some key terms:

- Columns (Fields): These define the attributes or characteristics. In our 'Students' table, 'Name' and 'Age' are columns.
- Rows (Records): Each row represents an individual entry or record in the table.

- Primary Key: A unique identifier for each record. For example, 'StudentID' could be the primary key in the 'Students' table.

The Relational Model: Connecting the Dots

In the realm of databases, relationships are everything. The relational model lets you link tables using keys, creating connections that mirror real-world scenarios. This is where the term 'relational database' comes from.

- Linking Tables Using Keys: Tables are connected using primary and foreign keys, enabling a seamless flow of information between them.
- Normalization: This is organizing data to reduce redundancy and improve integrity. It ensures that the data is stored logically and efficiently.

For example, in a school system, you might have tables for 'Students,' 'Courses,' and 'Instructors.' The relational model allows you to connect these tables, creating a coherent and interconnected system.

Real-Life Applications: The Importance of Structured Data

Why does all this matter? The answer lies in the tangible benefits that structured data offers:

- Faster Retrieval of Information: Like finding a book in a well-organized library, structured data enables quicker access to the needed information.
- Accuracy: By linking tables and normalizing data, you ensure accuracy and consistency across the entire database.
- Flexibility: Structured data allows for versatile querying, enabling complex analyses and insights.

Sketching a Scenario: School Management System

Let's bring all these concepts together by sketching a simple database schema for a school system:

- Students Table: Contains details about students, such as name, age, and enrolled courses.
- Courses Table: Houses course information, including course name and instructor.
- Instructors Table: Maintains records of instructors, their expertise, and the courses they teach.

These interconnected tables reflect real-world relationships between students, courses, and instructors. It's a system that encapsulates the essence of the school's operations, all managed and accessed through the elegant structure of databases and tables.

SQL Syntax

"Talk to your database as you would a person." If SQL were a language of conversation, its syntax would be its grammar. And just like any language, mastering grammar is essential to communicating clearly and effectively. Let's unravel the elegant structure of SQL syntax, the tool that lets us interact with databases and tables.

Speaking the Language: What is SQL?

SQL, or Structured Query Language, is a declarative language. It's like stating what you want for dinner without describing every step of cooking it. You tell the database what you need, and SQL handles the 'how.' It's a language designed to manage and manipulate relational databases, and its syntax serves as the rules of engagement.

The ABCs of SQL Syntax

Understanding SQL syntax is like learning the alphabet of a new language. Here's a breakdown of the essential elements:

- Commands: These are the verbs of SQL, the action words that tell the database what to do. Examples include SELECT, INSERT, DELETE, and UPDATE.
- Clauses: Clauses modify the commands to provide specific details. For example, the WHERE clause filters the data based on specific conditions.
- Expressions: These can produce either scalar values or tables of columns and rows. They combine operators and constants to create new values.
- Predicates: Predicates provide a way to compare values, often used within the WHERE clause to filter data.

Here's a simple example to illustrate:

```
SQL SELECT Name, Age FROM Students WHERE
Age > 18;
```

This query asks the database to retrieve the 'Name' and 'Age' columns from the 'Students' table for all students older than 18. It's a straightforward request, eloquently expressed in SQL syntax.

The Art of Conversation: Importance of SQL Syntax

Why bother with SQL syntax? The answer lies in the power of communication. SQL syntax allows you to:

- Interact with Databases: With the correct syntax, you can query, update, delete, or insert data. It's how you converse with the database.
- Manipulate Data: You can shape and transform data to suit your needs, whether for analysis, reporting, or decision-making.
- Ensure Accuracy: Proper syntax ensures that you're asking for exactly what you need, reducing errors and enhancing efficiency.

Just as clear speech fosters understanding in human conversation, precise SQL syntax facilitates smooth interaction with your data.

One Language, Many Dialects: Standardization of SQL

SQL speaks in many accents, and that's where the concept of standardization comes in. ANSI SQL is the standard version, but various database systems have their dialects. Think of it as the same language with regional variations.

For example, MySQL, SQL Server, and Oracle may slightly differ in interpreting specific commands. But the core of the language, the essence of the syntax, remains the same. It's like enjoying a favorite dish prepared slightly differently in various parts of the world. The ingredients are the same; the flavor is recognizable but has unique touches reflecting local culture.

A Glimpse into the Code: Examples of SQL Syntax

To make sense of SQL syntax, let's look at a few examples:

1. Retrieving Data:

```
SQL SELECT FirstName, LastName FROM
Employees;
```

This query fetches the first and last names of all employees.

2. Updating Data:

```
SQL UPDATE Products SET Price = Price *
1.10 WHERE Category = 'Electronics';
```

This query increases the price of all electronic products by 10%.

3. Deleting Data:

```
SQL DELETE FROM Orders WHERE Status =
'Cancelled';
```

This query removes all orders with a 'Cancelled' status.

Each line of code is a carefully crafted request, a question asked, or a task assigned. It's a conversation between you and the data, mediated by the eloquent structure of SQL syntax.

Basic Commands

> *"In every language, there are building blocks that form the foundation. In SQL, we have commands that allow us to converse with data, ask questions, give instructions, and even shape the world within our databases."*

The essence of SQL lies in its commands, each serving a unique purpose, enabling us to interact with our data like never before. Let's unravel these basic commands, akin to the tools in an artisan's toolkit, and explore how they shape, manipulate, and retrieve data.

SELECT: The Inquisitive Command

Definition: The SELECT command is the question-asker of SQL. It allows us to retrieve data from one or more tables, specifying exactly what we want to know.

Example:

```
SQL SELECT Name, Age FROM Employees WHERE
Department = 'HR';
```

This query requests the names and ages of all employees in the HR department.

Importance: Think of SELECT as the starting point for data exploration. Whether you're analyzing trends, generating reports, or satisfying your curiosity, SELECT forms the basis of data retrieval. The key opens the treasure trove of insights hidden in your data.

WHERE: The Filter of Precision

Definition: The WHERE clause in SQL acts as a filter, narrowing down the data based on specific conditions.

Example:

```
SQL SELECT Title FROM Books WHERE Genre = 'Science Fiction' AND Rating > 4;
```

This query finds all science fiction books with a rating greater than 4.

Importance: WHERE gives us control over the data we retrieve. It's like using a sieve to filter out the grains of information we need from a mound of data. Without WHERE, we would be overwhelmed by unnecessary details, losing sight of what truly matters.

INSERT INTO: The Creator of Content

Definition: INSERT INTO is the creator, the artist that adds new data to our tables.

Example:

```
SQL INSERT INTO Customers (FirstName, LastName, Age) VALUES ('John', 'Doe', 35);
```

This query inserts a new customer record into the Customer table.

Importance: Imagine a library without new books or a garden without fresh flowers. INSERT INTO breathes life into our databases, ensuring that they grow and evolve, reflecting the ever-changing reality of our world.

UPDATE: The Sculptor of Change

Definition: UPDATE is the sculptor that reshapes data, modifying existing records to keep them up-to-date.

Example:

```
SQL UPDATE Orders SET Status = 'Shipped'
WHERE OrderID = 1023;
```

This query updates the status of a specific order to 'Shipped.'

Importance: Change is inevitable, and UPDATE allows our databases to adapt. It ensures that our data remains relevant, accurate, and reflects current affairs.

CREATE: The Architect of Structure

Definition: CREATE is the architect that builds structures within our database. It can create new tables (CREATE TABLE) or entire databases (CREATE DATABASE).

Example - Table:

```
SQL CREATE TABLE Authors (AuthorID INT,
Name VARCHAR(50), Birthdate DATE);
```

Example - Database:

```
SQL CREATE DATABASE BookStore;
```

Importance: Without CREATE, there would be no data storage foundation. It's the first step in constructing the environment where our data resides, setting the stage for all subsequent interactions.

DELETE: The Eraser of Unwanted Data

Definition: DELETE is the eraser that removes data from our tables.

Example:

```
SQL DELETE FROM Cart WHERE UserID = 4567;
```

This query deletes all items from a specific user's cart.

Importance: Just as we prune a tree to encourage healthy growth, DELETE allows us to remove unnecessary or outdated data. It keeps our databases lean, efficient and focused on what truly matters.

ALTER: The Modifier of Existing Structure

Definition: ALTER is the modifier capable of changing the structure of existing tables (ALTER TABLE) or databases (ALTER DATABASE).

Example - Table:

```
SQL ALTER TABLE Employees ADD COLUMN Salary
DECIMAL(10, 2);
```

Example - Database:

```
SQL ALTER DATABASE BookStore SET RECOVERY
SIMPLE;
```

Importance: As our needs change, so must our databases. ALTER allows us to reshape and reconfigure, ensuring our structures align with our goals and requirements.

DROP: The Eliminator of Unwanted Structure

Definition: DROP is the eliminator, capable of removing entire tables or databases.

Example:

```
SQL DROP TABLE TemporaryData;
```

Importance: Sometimes, the best way to move forward is to let go of what's no longer needed. DROP allows us to do just that, removing unwanted structures and freeing up space for new growth.

Chapter 2

Hands-On Exercise: Writing Your First SQL Queries

Ah, the joy of transforming a concept into something tangible, something real! If you've been following along with me, we've traversed the landscapes of SQL, from understanding the architecture of databases to the power of manipulating data. But now, it's time to let you try your hand at it. You're more than ready, and I'll be right here to guide you.

Let's start by reflecting on our journey so far, shall we? Remember when we uncovered the secrets behind the SELECT command? Or how the WHERE clause can pinpoint the precise data we need? These weren't just abstract concepts; these were tools, and now it's time to wield them.

Embrace Your Inner Data Analyst

When I first learned about SQL, I remember feeling overwhelmed by the sheer complexity of it all. But let me share a secret: The magic lies in the execution. It's in crafting a query, the stroke of a keyboard, and the result displayed on your screen. That's when it all comes together.

The links provided earlier were valuable resources, but nothing compares to hands-on experience. So, let's get to it, shall we? You can complete a simple task using the SQL commands we've explored.

Task: Imagine you have a table named 'Employees' with columns 'ID,' 'Name,' 'Age,' 'Salary,' and 'Department.' Your task is to write a query that retrieves the Name and Age of all employees in the 'Marketing' department who earn more than $50,000.

Do you feel a bit lost? Don't worry; that's natural. Let's walk through the solution together, step by step.

Crafting the Query

1. Identifying What We Need

Before we even start typing, let's break down what we need. We want the Name and Age, so that's what we'll SELECT. We have specific conditions for the Department and Salary, so that's where our WHERE clause comes in.

2. Writing the SELECT Statement

Start with what you know. We need the Name and Age, so our query begins with:

```
SQL SELECT Name, Age
```

3. Adding the FROM Clause

We need to specify the table, don't we? So, let's add:

```
SQL FROM Employees
```

4. Crafting the WHERE Clause

Here's where things get interesting. We have two conditions, so we'll use:

```
SQL WHERE Department = 'Marketing' AND
Salary > 50000
```

5. Putting It All Together

Our final query looks like this:

```
SQL SELECT Name, Age FROM Employees WHERE
Department = 'Marketing' AND Salary > 50000
```

Simple, elegant, and powerful. That's SQL for you.

Encouragement and Assurance

This might feel daunting, but making mistakes is part of learning. Many talented data analysts stumble, fall, and then rise stronger. Your path is no different.

Try this query out on a sample database if you can. Play around with it. Modify the conditions, change the columns, and experiment with the structure. That's how learning happens.

And if you feel stuck? You're never alone. Revisit our discussions, refer to the resources, or take a moment to ponder. The answers are there, waiting to be discovered.

Segue: Preparing for the Next Adventure in SQL

"Success is not the result of spontaneous combustion. You must set yourself on fire."

— Arnold H. Glasow

And you, dear friend, have ignited the spark. You've lit the path toward a brighter future with your first SQL queries. You've taken the whispers of data and turned them into a symphony of understanding. You've seen how simple commands can unlock doors to information, insight, and wisdom.

Now, your curiosity stirs, and a new horizon beckons. What lies ahead is an exploration of the wild, untamed territories of data. You've proven your mettle courage and demonstrated your ability to wield the tools of SQL. The time has come to venture into the uncharted landscapes of data wrangling.

Data Wrangling: The Heart of Data Analytics

Imagine standing at the edge of a vast jungle. Trees tower above you, vines twist, and wind and the air is thick with mystery. This is the world of raw, unprocessed data. It's chaotic, messy, and teeming with life. But within that chaos lies the secret to powerful insights, and it's your mission to uncover them.

Data wrangling is akin to navigating this jungle. You'll learn to cut through the thicket, find the paths that lead to understanding, and forge new trails where none existed. You'll discover how to clean data, join tables, and craft more complex queries that reveal hidden patterns.

Cleaning Data: The Art of Refinement

Data is rarely perfect. In many ways, it's like a rough diamond, filled with impurities and flaws. But with the proper techniques, you can polish, shape, and turn it into something of immense value.

In the next chapter, we'll explore the methods of cleaning data. You'll learn how to handle missing values, correct inconsistencies, and ensure your data is ready for analysis. Think of it as sculpting, where every cut and every adjustment brings you closer to a masterpiece.

Joining Tables: Connecting the Dots

But what if one piece of information isn't enough? What if you need to see the connections between the relationships between different data sets? That's where joining tables comes into play.

You'll discover how to bring together disparate pieces of information, connect them meaningfully, and create a unified picture of your data landscape. It's like assembling a puzzle; every part fits just right, and the final image reveals something profound and beautiful.

More Complex Queries: Diving Deeper

And then there are the depths, the layers beneath the surface that hold the most profound secrets. With more complex queries, you'll learn to probe deeper, ask more nuanced questions, and uncover the subtleties that make data fascinating.

You'll explore subqueries, aggregations, and advanced filtering techniques that allow you to ask more intricate questions and find more precise answers. It's a journey into the heart of data, where the most exciting discoveries await.

Chapter 3
Data Wrangling

"The world is messy; the world is incomplete," observed the renowned statistician Nate Silver. It's a statement that rings true not just for life but also for the data that surrounds us. In data analytics, the messiness of real-world data can be both a challenge and an opportunity. Data wrangling techniques are like the compass and map that guides us through this chaotic terrain, helping to clean and shape data for better insights. Let's unravel the tangled threads of data wrangling and discover how SQL can be your tool for turning disorder into understanding.

The Messy Reality of Real-World Data

Data in the real world is rarely neat. It comes from various sources and formats, often with inconsistencies and errors. Whether missing values, duplicate records, or mismatched data types, the messiness reflects our complex and dynamic world.

But this messiness is not necessarily a bad thing. It's a sign of richness and diversity. It tells a story of human behavior, interactions, and the fluid nature of information. It's a puzzle waiting to be solved, a riddle longing to be unraveled.

Data Wrangling: Turning Chaos into Clarity

Data wrangling is the process of cleaning, structuring, and enriching raw data into a desired format for better decision-making in less time. It's like taking a lump of clay and molding it into a beautiful sculpture. The clay has potential but needs your hands, creativity, and understanding to turn it into art.

With its versatile and powerful syntax, SQL becomes your chisel and brush in this artistic process. It allows you to transform data, fix errors, fill gaps, and create a coherent picture from the chaos.

Integrating Data: Weaving the Threads Together

Once the data is clean, the next step is integration. It's about bringing together different strands of information and weaving them into a coherent fabric.

SQL offers powerful JOIN operations that allow you to combine data from different tables, creating new perspectives and insights. Whether INNER JOIN, LEFT JOIN, or

RIGHT JOIN, you can choose the method that best fits your needs and connects the dots meaningfully.

Transforming Data: Shaping the Sculpture

Transformation is shaping and molding the data into the desired form. It's where your creativity and understanding come into play.

With SQL, you can:

- Aggregate data: Summarize data using functions like SUM, COUNT, AVG, MAX, and MIN to create a higher-level view.
- Create new variables: Generate new columns using arithmetic or logical operations based on existing data.
- Filter and sort: Select specific data that meets certain criteria and arrange it to focus on what matters most.

Practical Exercise: Your Turn to Wrangle

Now it's your turn to try your hand at data wrangling. Below is a simple exercise to get you started:

1. Dataset: Assume you have a table of employees with columns employee_id, first_name, last_name, salary, and department.

2. Task 1: Write an SQL query to find the total salary for each department.
3. Task 2: Write an SQL query to find the highest-paid employee in each department.
4. Task 3: Write an SQL query to remove duplicate records based on employee_id.

Take your time to think, experiment, and discover the answers. Remember, there's no single "right" way to wrangle data. Your approach, your creativity, and your understanding will shape the outcome.

What is Data Wrangling?

"Raw materials are like wild animals; they are hard to tame." This quote by Sir Michael Moritz, a venture capitalist, may not be directly about data, but it strikingly captures the essence of what we're about to explore. In data analytics, raw data is akin to raw materials in industries like manufacturing or cooking. It's valuable but not very useful in its unrefined form. It's like a wild, untamed beast that needs to be understood, nurtured, and guided into something more meaningful. This is where data wrangling comes into play.

The Essence of Data Wrangling

Think of a chef in a bustling kitchen surrounded by many ingredients. Some are fresh, some are frozen, some are neatly chopped, while others are still whole. The chef's job is to transform these raw materials into a delightful meal. This culinary transformation is what data wrangling is to raw data.

Data wrangling, or munging, is the art of preparing and transforming raw data into a more suitable, clean format for analysis. It's about understanding the nature of the data, identifying its quirks and inconsistencies, and applying the proper techniques to prepare it for the principal course analysis.

The Necessity of Data Wrangling

Why is data wrangling necessary? Imagine building a house with different sizes, shapes, and quality bricks. It's chaotic and inefficient, and the final structure would be shaky. Inconsistent data is similar. It causes confusion, leads to errors, and can significantly hamper the data analysis.

Real-world challenges like missing data, duplicates, and inconsistencies are not merely technical glitches but reflections of the complexity of human behavior and the dynamism of information. Data wrangling is the bridge that connects this messy reality to the structured world of data analytics. It's about finding the harmony in the chaos, the signal in the noise.

The Importance of Data Wrangling in Data Analysis

Regarding data analysis, the significance of data wrangling cannot be overstated. It's said that data scientists spend 80% of their time wrangling data based on the 80/20 rule. It's a staggering figure that highlights this phase's crucial in the overall process. Interestingly, other sources put this figure at around 45%, but the point remains that data wrangling takes up a significant portion of the entire data analysis lifecycle.

What does this mean for you as a budding data analyst or a professional dealing with data? It implies that mastering data wrangling is not just a technical skill; it's a vital part of your toolkit that empowers you to unlock the hidden potentials of data.

Techniques for Data Cleaning in SQL

"I have striven not to laugh at human actions, not to weep at them, nor to hate them, but to understand them." This quote by Baruch Spinoza might seem odd in the context of data cleaning, but it embodies a profound truth. Understanding is at the heart of any transformation, including transforming raw, messy data into clean, insightful information. This section will explore the specific methods used to clean data using SQL, allowing us to understand and refine our data's nuances for analysis.

Chapter 3

Understanding the Data Cleaning Process

Data cleaning is not a singular action; it's a series of thoughtful processes, each addressing a unique challenge in the raw data. From identifying and handling missing data removing duplicates, to fixing inconsistent entries, data cleaning is about recognizing the imperfections and applying the right solutions.

Missing Data

Missing data is like a puzzle with pieces absent. It can distort the picture and lead to inaccurate conclusions. In SQL, we can identify missing data using the IS NULL condition. For example:

```
SQL SELECT * FROM students WHERE grade IS
NULL;
```

This query retrieves all the rows from the students table where the grade column has missing values. We can then decide whether to fill these missing values with a default value, interpolate them based on other data, or exclude them from analysis.

Removing Duplicates

Duplicates are like echoes; they create unnecessary noise. SQL provides a simple way to remove duplicates using the DISTINCT keyword. Here's an example:

```
SQL SELECT DISTINCT name FROM employees;
```

This query returns all the unique names from the employees table, effectively removing duplicate names.

Fixing Inconsistent Entries

Inconsistent entries are like mispronounced words; they can confuse and mislead. SQL offers various functions to handle these inconsistencies. For example, using the UPPER function, we can standardize text entries:

```
SQL SELECT UPPER(city) FROM locations;
```

This query converts all the city values in the locations table to uppercase, ensuring uniformity.

SQL Queries: The Magic Wands of Data Cleaning

SQL queries are not mere lines of code; they're magic wands that transform the raw, unruly data into a structured, meaningful form. Let's delve into some examples that show-case this transformation:

Handling Missing Values

Consider a table of orders with some missing values in the shipping_date column. We can fill these missing values with a default date using the following query:

```
SQL UPDATE orders SET shipping_date =
'2021-01-01' WHERE shipping_date IS NULL;
```

This query updates all the rows with missing shipping_date and sets them to a default date.

Removing Specific Duplicates

Imagine a products table with duplicate product names. We can remove specific duplicates based on certain conditions using a query like:

```
SQL DELETE FROM products WHERE product_id
NOT IN (SELECT MIN(product_id) FROM prod-
ucts GROUP BY product_name);
```

This query keeps the product with the minimum product_id for each product_name and deletes the rest, effectively removing the duplicates.

SQL Functions: The Fine Brushes of Data Cleaning

SQL functions are like fine brushes in an artist's toolkit. They allow us to make detailed, precise changes to the data. Here's a glimpse of some powerful SQL functions used for data cleaning:

TRIM

The TRIM function removes leading and trailing spaces from a string:

```
SQL SELECT TRIM(product_name) FROM products;
```

UPPER/LOWER

We've already seen how the UPPER function can standardize text. The LOWER function does the same but converts text to lowercase:

```
SQL SELECT LOWER(email) FROM customers;
```

REPLACE

The REPLACE function allows us to replace specific characters or strings within a text:

```
SQL SELECT REPLACE(address, 'St.', 'Street') FROM suppliers;
```

Methods & Best Practices for Data Consistency

"Consistency matters the most in triggering something important to your life." Abdul Rauf's words resonate with more than just life's philosophies. In the world of data analysis, the concept of consistency is fundamental and often a hidden challenge that many data professionals face. Let's explore this critical aspect of data, why it's essential, and how SQL acts as a guardian to maintain data consistency.

The Essence of Data Consistency

Data consistency is more than a technical term; it's the assurance that your data speaks the truth and sings the same tune across different platforms and queries. It means the data is reliable, harmonious, and contradictions-free. Imagine a symphony where each instrument plays harmoniously, creating a beautiful melody. Data consistency aims to achieve that in the vast orchestra of information.

Why is it essential in data analysis? Data can become a cacophony of mismatched notes without consistency, leading to incorrect interpretations and flawed decisions. Inconsistent data is like a mirage; it appears real but leads you astray.

The Bumps in the Road: Common Issues Affecting Data Consistency

The path to data consistency is not always smooth. There are bumps, twists, and turns that can derail the process. Some common issues affecting data consistency include:

1. Duplicate Data: Like an annoying echo, duplicate data creates confusion. The presence of the same information in multiple places leads to inconsistency.
2. Inaccurate Data: Inaccurate data is like a wrong note in a melody; it stands out and disrupts the flow. The presence of incorrect or outdated information doesn't reflect the current reality.
3. Mismatched Formats: Different formats can create chaos. Imagine trying to blend oil and water; they don't mix. Mismatched structures lead to a lack of uniformity and create barriers in data analysis.
4. Lack of Data Validation: Data can grow uncontrollably and in all directions like wild weeds. Lack of validation allows incorrect data to enter the system unchecked.

SQL: The Guardian of Data Consistency

With its robust features and commands, SQL acts as a guardian to ensure data consistency. It's like the orchestra's conductor, guiding each instrument to play harmoniously. Here's how SQL helps in maintaining data consistency:

1. Data Integrity Constraints: SQL offers constraints like PRIMARY KEY, UNIQUE, and CHECK that enforce specific rules on data, preventing inconsistencies.

2. Data Validation: SQL allows the creation of triggers and stored procedures to validate data before it enters the system. It's like a gatekeeper, ensuring that only the correct data passes through.
3. Consistency Checks: SQL provides tools to run consistency checks, ensuring that the data adheres to the defined rules and standards.

For example, to ensure that the salary field in an employees table is never negative, you can use a CHECK constraint:

```
SQL CREATE TABLE employees ( employee_id
INT PRIMARY KEY, name VARCHAR(50), salary
DECIMAL CHECK (salary > 0) );
```

This constraint ensures that the salary value is always positive, maintaining consistency.

Best Practices: The Compass to Navigate the Path of Consistency

Maintaining data consistency is like navigating through a dense forest; you need a compass, a guide to keep you on the right path. Here are some best practices that act as that compass:

1. Use Data Validation Rules: Implementing strict validation rules ensures that the data entering the system meets the required standards.

2. Implement Consistency Checks: Regular consistency checks help identify and correct any inconsistencies that might have crept into the system.

3. Maintain a Standard Format: Adhering to a standard format across all data sources creates uniformity and makes analysis more accessible.

4. Avoid Manual Entry Where Possible: Automation reduces human error, one of the primary sources of inconsistency.

These practices are like the cardinal directions on a compass, guiding you through the complex landscape of data consistency.

Hands-On Exercise: Cleaning a Sample Dataset

"A clean canvas leads to a beautiful painting." This quote by an anonymous artist may not directly talk about data, but it beautifully captures the essence of data cleaning. Just like a clean canvas enables the painter to create art, a clean dataset sets the stage for profound insights and intelligent decision-making. In this exercise, we'll roll our sleeves, dive into a sample dataset, and get our hands dirty with real-world data cleaning.

Chapter 3

Here's Your Challenge: A Sample Dataset with Inconsistencies

Take a look at this dataset. It's filled with various inconsistencies and issues. There are missing values, duplicate entries, inaccurate data, and mismatched formats. It's like a puzzle with pieces all jumbled up, waiting for you to solve.

Step by Step: Guiding You Through the Cleaning Process

Let's begin the cleaning process. Don't worry; I'll be right here with you, guiding you through each step. The dataset may look messy now, but we can make it a valuable asset with a little effort.

1. Identify and Remove Duplicates: Like removing weeds from a garden, start by identifying and eliminating duplicate entries. It will make the data more concise and consistent.
2. Handle Missing Values: Missing values are like gaps in a story, making it hard to understand. Fill these gaps by either removing the rows with missing values or replacing them with appropriate values, such as the mean or median of the column.
3. Correct Inaccurate Data: Inaccurate data can lead you down the wrong path. Correct or remove any incorrect data, verifying it against reliable sources if necessary.

4. Standardize Formats: Having data in different formats is like trying to fit square pegs into round holes. Standardize the formats across columns to ensure uniformity.

5. Apply Data Validation Rules: Set up validation rules to ensure that the data meets specific criteria, like using a CHECK constraint in SQL to enforce conditions.

6. Document the Cleaning Process: Just as a chef notes a recipe, document the cleaning process. It will serve as a guide for future reference and ensure consistency across different datasets.

Here's an example SQL query that shows how to remove duplicates from a table named employees:

```
SQL DELETE FROM employees WHERE employee_id
NOT IN ( SELECT MIN(employee_id) FROM
employees GROUP BY name, position, salary
);
```

This query identifies duplicates based on the name, position, and salary columns and removes them, retaining only one instance of each duplicate record.

The Importance of Each Technique

Every technique we've used in this exercise has its importance, just like every brush stroke contributes to the final painting.

- Removing Duplicates: Enhances accuracy by ensuring that each record is unique.
- Handling Missing Values: Ensures completeness so you don't miss out on insights.
- Correcting Inaccurate Data: Builds trust by making the data reliable.
- Standardizing Formats: Facilitates analysis by creating uniformity across the data.
- Applying Validation Rules: Acts as a preventive measure to maintain consistency in the future.

Encouragement: Mastery Comes with Practice

You've done it! You've cleaned the dataset, transforming it from a jumbled mess into a valuable asset. But remember, mastery comes with practice. Don't hesitate to experiment with different techniques and explore various datasets. Your growth as a data professional lies in these hands-on experiences.

I understand the apprehension, the fear of messing up. It's natural to feel overwhelmed, especially when handling real-world data. But believe in yourself. Embrace the challenges, learn from mistakes, and keep honing your skills. I'm here with you, cheering you on, confident that you've got what it takes to excel in the fascinating world of data analytics.

Segue:

"Two minds are better than one," goes the old saying. But what about data? Could it be that two datasets are better than one? Indeed, the richness and depth of insights often lie in integrating various data sources. It's like weaving distinct threads into a fabric that tells a story that might remain concealed if each thread were examined individually.

Bridging the Gaps: The Art of Data Integration

The world of data is vast, complex, and multifaceted. Datasets from different sources may contain puzzle pieces that, when combined, reveal patterns and connections that might otherwise go unnoticed. It's akin to hearing different instruments in an orchestra and recognizing the beauty when they play together, harmonizing into a cohesive melody. Oops! I promised not to use musical metaphors, didn't I? Let's get back to our data world.

Data integration is about bridging the gaps between diverse datasets. Imagine having customer data in one system, sales data in another, and marketing data in another. Individually, they tell a part of the story. Together, they provide a holistic view of the business.

Chapter 3

Challenges: A Path Strewn with Obstacles

But it's not all smooth sailing. Combining data from various sources comes with its challenges. Formats may vary, inconsistencies lurk, and matching records across systems could be daunting. Have you ever tried fitting pieces from different puzzles together? It's not easy, and neither is data integration.

The challenges are real, but so are the rewards. Overcoming these obstacles unlocks deeper insights and a broader understanding of the interconnected nature of data. It's worth the effort but requires skill, diligence, and a keen eye for detail.

Techniques: Tools in Your Arsenal

There are several techniques and tools to aid in data integration:

- Joining Tables: SQL provides powerful joining capabilities that combine tables based on common attributes. It's like finding familiar friends in different social circles and connecting them.
- Data Warehousing: Creating a centralized repository that houses data from various sources enables a unified view. Think of it as a library with books from different genres, all under one roof.

- ETL Processes: Extract, Transform, and Load (ETL) processes are the backbone of data integration. They enable the extraction of data from various sources, transformation into a consistent format, and loading into a target system. Picture a chef gathering ingredients from different places, preparing them, and cooking a delicious meal.

- APIs and Connectors: Modern technologies offer APIs and connectors that facilitate seamless integration between disparate systems. It's as if you had a universal adapter that fits all plugs.

Real-World Applications: Where Integration Shines

Data integration finds application across industries, from healthcare to finance, marketing to technology. In healthcare, patient records from different systems can be integrated to provide comprehensive care. Integrating market data, customer profiles, and transaction history enables robust risk analysis in finance.

It's not just about business; data integration has societal impacts, too. Integrating climate data from various sources can lead to better weather predictions and disaster management. It's about connecting dots, building bridges, and unveiling insights that make a difference.

Chapter 3

Your Turn: Explore and Experiment

The realm of data integration awaits your exploration. Start with simple joins, experiment with ETL processes, explore data warehousing, and delve into APIs and connectors. Every step will unveil a new layer of understanding, connection, and insight.

And as you explore, remember, you're not just manipulating numbers and strings; you're weaving a narrative, telling a story, unlocking secrets hidden in the data. It's a fascinating world with challenges, rewards, complexities, and simplicities waiting for you to discover.

So here's to the integrators, the bridge builders, the dot connectors. May your efforts lead to wisdom, struggles to mastery, and curiosity to endless discoveries. The next chapter awaits, where we'll take this integration to even greater heights. But for now, let's pause and reflect on the beauty of interconnected data, which lies in unity, coherence, and the graceful dance of... oh, there I go again with the forbidden metaphors! Let's say the beauty of integration.

Chapter 4
Data Integration

> *"Our world generates 2.5 quintillion bytes of data every day. Yes, you read that right—2.5 quintillion! The sheer magnitude of this number is beyond human comprehension. But what's even more astonishing is the diversity of this data—numbers, text, images, videos—all swirling around in a chaotic dance of information. How do we make sense of it all? How do we integrate this vast ocean of data into something coherent and meaningful?"*

D ata integration is not merely an academic exercise; it's a real-world challenge that touches every aspect of our lives. Whether you're a business analyst looking to combine sales data from various channels, a researcher trying to integrate experimental data with existing litera-ture, or a healthcare professional aiming to consolidate patient information from multiple sources, the task of data integration looms large. And as the data grows in volume, variety, and velocity, this task becomes even more daunting.

Why Data Integration Matters

"Imagine being a detective on the trail of a complex mystery. You have clues scattered across various locations: a fingerprint in a dusty room, a strand of hair at the crime scene, an eyewitness account in a crowded street. Individually, these clues might seem insignificant, but together, they form a coherent picture, leading you to the truth. Data integration is much the same. It's about piecing together fragments of information from different sources to reveal something greater. It's about connecting the dots."

I remember working with a large e-commerce company struggling with data fragmentation. They had customer data spread across marketing, sales, and customer service departments. Marketing had information about customer preferences and browsing habits. Sales had data on purchases, preferences, and buying patterns. Customer service had records of complaints, queries, and feedback.

The company wanted to create a personalized marketing campaign for their customers. They would need a complete view of each customer, including their preferences, purchases, and interactions with customer service. Without integrating this data, they would only have a fragmented picture, like trying to complete a puzzle with missing pieces.

The Essence of Data Integration

Data integration combines data from various sources into a unified view. It's like weaving a tapestry with threads from different fabrics, each representing information from another source. The final tapestry is rich, colorful, and coherent, offering insights that were not apparent when the threads were viewed separately.

But why does data integration matter so much? Why not just analyze the data in its isolated silos?

Imagine you're a healthcare provider trying to create a comprehensive health profile for a patient. You have medical records from various hospitals, lab reports from different laboratories, and prescriptions from multiple pharmacies. Analyzing these separately would give you an incomplete picture of the patient's health. Integrating this data, on the other hand, would allow you to see patterns, correlations, and connections that would otherwise remain hidden.

Or consider a financial analyst trying to assess the performance of a multinational corporation. They would need data from various departments, subsidiaries, and countries. Analyzing this data in isolation would be like trying to understand a symphony by listening to each instrument separately. Only by integrating the data can the analyst see the complete performance of the company.

The Role of Data Integration in Providing a Unified View

The strength of data integration lies in its ability to provide a unified view of data gathered from various sources. It breaks down the walls between silos, allowing data to flow freely and coalesce into something more significant.

Let's go back to the e-commerce company I mentioned earlier. By integrating customer data from marketing, sales, and customer service, they created personalized marketing campaigns that resonated with their customers. They could see which products a customer had viewed but not purchased, which they had bought in the past, and what issues they had raised with customer service. This integrated view allowed them to craft marketing messages that were highly relevant and targeted, leading to increased sales and customer satisfaction.

But it's not just about marketing. Data integration plays a vital role in various domains:

- In Healthcare: Integrating patient records from various providers enables comprehensive care, ensuring doctors have all the information they need to make informed decisions.

- In Finance: Combining data from various financial systems allows for more accurate risk assessment, fraud detection, and investment analysis.
- In Manufacturing: Integrating data from production, sales, and supply chain systems enables efficient inventory management, production planning, and demand forecasting.

The possibilities are endless, and the benefits are profound. Data integration is not just a technical process; it's a strategic enabler, unlocking insights, enhancing decision-making, and driving value.

The Journey Towards Data Integration: A Practical Exercise

Let's embark on a practical exercise to illustrate the power of data integration. Suppose you're a business analyst working for a retail chain. You have sales data from various stores, inventory data from warehouses, and customer feedback from online platforms. Your task is to integrate this data to analyze sales trends, manage inventory, and improve customer satisfaction.

1. Identify the Sources: Start by identifying your data sources. You have sales records from stores, inventory levels from warehouses, and customer feedback from online platforms.

2. Define the Common Keys: Determine the common keys allowing you to join the data. These could be product IDs, store locations, or customer IDs.
3. Clean and Transform: Clean the data, remove duplicates, handle missing values, and transform it into a consistent format.
4. Integrate the Data: Use tools like SQL to join the data on the common keys, creating a unified view of sales, inventory, and customer feedback.
5. Analyze and Act: With the data integrated, analyze the sales trends, identify inventory bottlenecks, and understand customer feedback. Use these insights to make informed decisions, such as adjusting inventory levels, launching targeted promotions, or improving customer service.

Why Data Integration is Important

 "The whole is greater than the sum of its parts."

— Aristotle

This ancient wisdom holds in data analytics and speaks directly to the core of data integration. Data integration is not merely gluing together different data pieces; it's about transforming disjointed information into a powerful asset that enhances our understanding and decision-making. Specifically, it plays a pivotal role in three key areas: business decision-making, operational efficiency, and improving customer experience.

Business Decision-Making: The Clarity of Integrated Insight

Remember when business decisions were made based on gut feelings or limited information? Those days are long gone. Decisions must be data-driven, informed, and strategic in the current business landscape.

But what happens when your data is scattered across various systems and formats? You end up with a blurred vision that hampers your ability to make insightful decisions.

This is where data integration comes to the rescue. Unifying data from various sources coherently provides a clear and comprehensive view that enables informed decision-making.

Think about a marketing manager trying to allocate a budget across different channels. Without integrated data, they might have information about website traffic, social media engagement, and email open rates in separate silos. But by incorporating this data, they can see how customers are moving through the entire marketing funnel, identify the most effective channels, and allocate budget accordingly.

Data integration also facilitates collaboration across different departments. Sales, marketing, and finance can work together, leveraging the same integrated data to align their strategies and goals. This collaborative approach leads to more consistent and effective business decisions that drive growth and success.

Operational Efficiency: The Engine of Streamlined Processes

In a rapidly changing business environment, operational efficiency is not just a nice to have; it's necessary. Wasting time and resources on manual data handling, inconsistency, and error-prone processes can be a significant roadblock to success.

Data integration streamlines operations by automating data collection, transformation, and consolidation. It eliminates manual errors, ensures data consistency, and frees up valuable time that can be invested in more strategic activities.

Imagine an e-commerce business dealing with inventory management. They have multiple suppliers, warehouses, and sales channels. Keeping track of inventory levels, order processing, and shipping across these different systems can be a nightmare without data integration.

By integrating data across suppliers, warehouses, and sales channels, the business can have a real-time view of inventory levels, automate order processing, and optimize shipping routes. The result is efficiency, increased accuracy, agility, and responsiveness to market demands.

Improving Customer Experience: The Heartbeat of Success

In today's competitive market, customer experience is often the differentiating factor that sets a business apart. Understanding customer needs, preferences, and behavior is crucial to delivering personalized experiences that delight and retain customers.

Data integration is the key to unlocking this understanding. It combines customer interactions and transactions across various touchpoints, such as websites, mobile apps, call centers, and physical stores. This integrated view provides a 360-degree understanding of the customer, allowing businesses to tailor products, services, and interactions to individual needs.

Consider a bank trying to provide personalized financial advice to its customers. The bank can completely understand a customer's financial situation by integrating data from savings accounts, loans, credit cards, and investment portfolios. This understanding enables them to offer personalized advice, products, and services that resonate with individual needs and goals.

Furthermore, integrated customer data enables real-time responsiveness. If a customer raises a complaint on social media, the customer service team can immediately access their entire history, understand the context, and address the issue promptly and effectively.

How SQL Comes into Play in Data Integration

> *"The future belongs to those who see possibilities before they become obvious."*
>
> — John Sculley

Ah, SQL! The heart of data manipulation, the soul of data retrieval, and the intricate language that allows us to converse with data. But how does SQL fit into the grand scheme of data integration, and why is it such a pivotal tool in today's data-driven world? Let's embark on this exciting exploration together.

SQL: The Conductor of Data Symphony

Data in isolation is like musical notes scattered on a page without a melody. SQL comes into play by combining these disparate notes and orchestrating a harmonious symphony. SQL is a conductor in data integration, directing data flow, ensuring consistency, and transforming raw information into meaningful insights.

Chapter 4

Bringing Together Varied Sources

In organizations large and small, data is often scattered across different databases, spreadsheets, applications, and even physical documents. SQL allows us to tap into these various sources, regardless of their format or location, and combine them into a unified view.

Think of a healthcare provider managing patient records. They might have demographic data in one database, medical histories in another, and lab results in a separate application. SQL enables them to query these diverse sources, integrate the data, and present a comprehensive view of each patient.

Transforming and Cleaning Data

One of the significant challenges in data integration is the inconsistency and incompleteness of data. Different sources might use other conventions, formats, or terminologies. SQL provides the tools to transform and clean this data, ensuring consistency and usability.

Imagine you're working with sales data from different regions, and each region reports currency in its local format. With SQL, you can convert all these different currencies into a standard format, aligning the data and making it comparable.

Enabling Complex Queries and Analysis

SQL's power lies in its ability to conduct complex queries and analyses. Once the data is integrated, SQL allows you to explore it, ask intricate questions, and derive insights that inform decision-making.

Consider a retail business trying to understand customer buying behavior. By integrating data from online sales, in-store purchases, customer service interactions, and social media engagement, SQL enables them to query this combined data set, identify patterns, and tailor marketing strategies accordingly.

Facilitating Real-Time Integration

In today's fast-paced environment, real-time data integration is not a luxury; it's a necessity. SQL facilitates this real-time integration, allowing businesses to have up-to-the-minute views of their operations, markets, and customers.

Imagine a stock trading platform that needs to integrate real-time stock prices, historical trends, and customer portfolios. SQL enables them to do this, providing traders with the timely information they need to make informed investment decisions.

Hands-On Exercise: Integrating Multiple Datasets

In the previous sections, we've delved into the theory of data integration and how SQL plays a vital role in this essential process. Now, it's time to roll up our sleeves and get hands-on. This exercise sharpens your axe, preparing you to tackle real-world data integration challenges.

Sample Datasets with Inconsistencies

To make this exercise as realistic as possible, we'll work with multiple small sample datasets, each containing some inconsistencies. These inconsistencies could be in terms of formatting, missing values, or variations in terminology.

Dataset 1: Customer Information

This dataset contains basic customer information, including Name, Age, Email, and Country. There might be inconsistencies in how countries are named (e.g., USA vs. United States).

Dataset 2: Order Details

This dataset includes details of customers' orders, such as Order ID, Product, Quantity, and Price. The product names might vary slightly between records.

Dataset 3: Product Catalog

This dataset provides a catalog of products with attributes like Product ID, Product Name, Category, and

Manufacturer. There might be variations in the category names.

Integrating These Datasets Using SQL

Now, let's embark on the exciting process of integrating these datasets using SQL. We'll guide you through each step, explaining our actions and why.

Step 1: Data Cleaning

Before integrating the datasets, we need to clean and make them consistent. We can standardize the data in SQL with TRIM, UPPER, and REPLACE.

For example, to ensure that all country names in the Customer Information dataset are consistent, we can use:

```
SQL UPDATE customers SET country = UPPER(TRIM(country));
```

Step 2: Data Transformation

Next, we'll transform the data to make it compatible across datasets. This might involve converting data types or mapping values.

For instance, we might need to map the product names in the Order Details dataset to the Product IDs in the Product Catalog. We can achieve this with a JOIN operation:

```
SQL UPDATE orders SET orders.product_id =
catalog.product_id FROM product_catalog AS
catalog WHERE orders.product_name = cata-
log.product_name;
```

Step 3: Data Integration

Now that the datasets are cleaned and transformed, we can integrate them into a comprehensive view.

We'll create a new table that combines customer information, order details, and product catalog:

```
SQL SELECT c.name, c.email, o.order_id,
p.product_name, o.quantity FROM customers
AS c JOIN orders AS o ON c.customer_id =
o.customer_id JOIN product_catalog AS p ON
o.product_id = p.product_id;
```

This SQL query gives us an integrated view of the customers, their orders, and the products they purchased.

Highlighting the Improvement

After the integration, you'll notice a significant improvement in the overall quality and comprehensiveness of the data. The inconsistencies are resolved, the information is aligned, and we have a unified view, allowing meaningful analysis and insights.

Segue:

"The goal is to turn data into information, and information into insight."

— Carly Fiorina

A sense of accomplishment fills the room as you glance back at the code you've just written. The datasets, once messy and disparate, now sit harmoniously in a unified structure. The inconsistencies have been ironed out, and the data is ready to reveal its secrets. It feels like you've unlocked a treasure chest, and now it's time to explore the riches within.

But what's next? You've integrated the data, but the adventure has just begun. Now it's time to delve deeper into SQL and harness its power to explore, analyze, and visualize the data.

Chapter 5
Advanced SQL Queries for Data Exploration

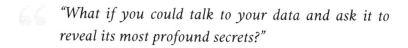 *"What if you could talk to your data and ask it to reveal its most profound secrets?"*

I t's a compelling proposition. The idea is that within the rows and columns of your datasets lie answers to questions you haven't even thought to ask. The notion is that your data is not just a collection of numbers and strings but a living, breathing entity waiting to tell you a story. This is where advanced SQL queries come into play, giving you the tools to engage in a meaningful dialogue with your data uncovering insights that can transform your decision-making process.

Exploring the Depths of Your Data

I recall a time when I was faced with a particularly intricate dataset. On the surface, it seemed ordinary, but I knew that hidden within its structure were insights that could reshape our understanding of a specific business problem. The challenge was to find the right tools to dig deep into the data.

Advanced SQL queries became my ally, allowing me to navigate the complex maze of information, uncovering patterns, trends, and connections that were previously obscured.

Crafting Queries that Speak to Your Data

Imagine being able to ask your data a question as if it were a person sitting across the table from you. What would you ask? What secrets would you seek to uncover?

Advanced SQL queries empower you to do just that. You can use subqueries, common table expressions (CTEs), and window functions to create conversations with your data, teasing out the subtleties that lead to proper understanding.

For example, consider a query that identifies the top-performing salesperson in each region:

```
SQL WITH RegionalTopSales AS ( SELECT
region, salesperson, RANK() OVER
(PARTITION BY region ORDER BY sales DESC)
AS rank FROM sales_data ) SELECT region,
salesperson, sales FROM RegionalTopSales
WHERE rank = 1;
```

This query uses a window function to rank salespeople within each region and then filters the results to find the top performer. It's like asking the data, "Who is the best in each region?" and receiving a clear and concise answer.

Exercises: Query Building Challenges

Now, it's your turn to engage with your data. Try these exercises to hone your advanced SQL skills:

1. Find the Average Sale by Category: Write a query that calculates the average sale for each product category in your database.
2. Identify the Most Profitable Customers: Use a subquery to identify the customers who have generated the highest profits for your company.
3. Analyze Seasonal Trends: Utilize window functions to analyze how sales trends change with the seasons.

Remember, the more you practice, the more fluent you become in speaking the language of SQL.

The Art of Data Visualization with SQL

Once you've explored your data, the next step is to present it in a way others can understand. This is where data visualization comes into play.

Advanced SQL queries provide you with the raw material for compelling visualizations. You can create charts, graphs, and dashboards that tell a story by shaping the data precisely.

Consider the thrill of turning a complex SQL result set into a colorful line chart that tracks sales trends over time. Or the satisfaction of building a heat map highlighting regions where your products are most popular. These visual representations add layers of understanding that go beyond mere numbers.

Advanced SQL Queries

 "The true delight is in the finding out rather than in the knowing."

— Isaac Asimov

The Difference Between Basic and Advanced SQL Queries

Basic SQL queries, those familiar companions, have served you well. They've allowed you to navigate your data, extract information, filter results, and create simple insights. But there's more, isn't there? A nagging curiosity that whispers in your ear, urging you to go further, dig deeper, and ask more complex questions. That's where advanced SQL queries come into play.

Advanced SQL queries are like unlocking a hidden chamber in a treasure-laden castle. They open up new possibilities, enabling you to perform intricate analyses, combine data from various sources, and uncover insights that were previously out of reach.

These are not mere extensions of the basic queries you've been using. They are powerful tools that allow you to perform JOINs, subqueries, and set operations. While basic queries might enable you to ask, "What are the sales figures for Product A?" advanced queries empower you to ask, "How do the sales of Product A compare across different regions, and what are the underlying patterns that might influence these sales?"

A Brief Introduction to the Types of Advanced SQL Queries

The world of advanced SQL queries is rich and varied, and the techniques you'll learn in this chapter are like precious gems waiting to be discovered.

- JOINs: These allow you to combine data from different tables, creating a unified view that facilitates more nuanced analyses.
- Subqueries: Think of subqueries as questions within questions. They enable you to use the result of one query as the input for another, opening up endless possibilities for exploration.

- Set Operations: These include UNION, INTERSECT, and EXCEPT, which allow you to combine, compare, or contrast the results of different queries.

Here's a tantalizing glimpse of what you can achieve with a JOIN:

```sql
SQL SELECT customers.name, orders.order_date FROM customers JOIN orders ON customers.customer_id = orders.customer_id WHERE orders.order_date > '2021-01-01';
```

This query connects the customers and orders tables, allowing you to see customer names alongside their order dates for orders placed after a specific date. It's like conversing with your data, asking it to reveal previously hidden connections.

The Potential of Advanced Queries to Reveal Deeper Insights

The beauty of advanced SQL queries lies in their ability to transform data into knowledge. They allow you to ask profound questions, engage with your data on a deeper level, and uncover patterns, trends, and relationships that basic queries might miss.

Perhaps you're wondering how a particular product's sales are influenced by seasonal trends or how customer behavior changes in response to specific marketing campaigns. Advanced queries allow you to explore these complexities, turning raw data into actionable insights.

Confidence in Handling Advanced Queries

I know the term "advanced" might sound intimidating. But let me assure you, these queries are not as fearsome as they may seem. With practice, patience, and persistence, you'll find that you can wield them with confidence, unraveling the mysteries of your data with the skill of a seasoned detective.

Remember, the goal here is not merely to learn commands or memorize syntax. It's to cultivate a mindset, a way of thinking that sees beyond the surface of the data and seeks to understand the stories it has to tell.

Data Exploration

 "Give me six hours to chop down a tree, and I will spend the first four sharpening the axe."

— Abraham Lincoln

Exploration is akin to sharpening your axe before felling the tree in data analytics. It's about preparing, understanding, and aligning your data to answer complex questions and unearth hidden patterns. SQL's advanced capabilities, such as JOINs, subqueries, and set operations, become your most potent allies in this phase.

A Closer Look at Subqueries

Subqueries, or nested queries, are intriguing, aren't they? They're like those Russian nesting dolls, one question hiding inside another, each revealing something new and unexpected. A subquery is a query within another SQL query that retrieves data that will be passed to the main query as a condition to restrict further the data retrieved.

Types of Subqueries

1. **Scalar Subquery:** This returns a single value. It's concise and precise, answering precise questions. For example:

```
SQL SELECT name FROM employees WHERE salary
> (SELECT AVG(salary) FROM employees);
```

Here, the subquery finds the average salary, and the main query uses that information to find all employees earning more than the average.

2. **Correlated Subquery:** This is an intertwined query where the subquery refers back to the outer query. Imagine it as a dance between two queries, each responding to the other.

```
SQL SELECT name, salary FROM employees e1
WHERE salary > (SELECT AVG(salary) FROM
```

employees e2 WHERE e1.department = e2.department);

In this example, the subquery calculates the average salary for each department, and the main query compares individual salaries to these averages.

3. Multiple Row Subquery: Unlike scalar subqueries, these return multiple rows. They often use operators like IN, ANY, or ALL.

SQL SELECT name FROM products WHERE category_id IN (SELECT category_id FROM categories WHERE name = 'Electronics');

Here, the subquery finds the category IDs for electronics, and the main query retrieves all products in those categories.

Set Operations in SQL

Set operations are your mathematical friends in SQL. They help you combine the results of two or more SELECT queries into a single result set.

1. UNION: Consider adding two result sets together without duplicate values.

SQL SELECT name FROM table1 UNION SELECT name FROM table2;

2. **INTERSECT:** This helps you find common elements in two result sets.

```
SQL SELECT name FROM table1 INTERSECT
SELECT name FROM table2;
```

3. **EXCEPT:** Imagine you want to find what's in one result set but not the other; that's where EXCEPT comes in handy.

```
SQL SELECT name FROM table1 EXCEPT SELECT
name FROM table2;
```

Revisiting JOINs for Exploration

Recall our earlier discussion on JOINs. They are like bridges connecting different tables, allowing you to access information across your database. Whether it's LEFT, RIGHT, FULL, CROSS, or SELF JOIN, each has unique use cases and capabilities, enabling you to craft intricate queries that unveil the stories within your data.

Practical Use-Cases and Exploration

Data exploration with SQL is like a detective novel filled with twists and turns, where queries act as clues leading to revelations. Here are some practical use cases:

- Subqueries: They are incredibly flexible and can be used to filter data, find discrepancies, or calculate aggregate values for specific conditions.

- Set Operations: These are perfect for comparing two datasets, finding common elements, or identifying differences.
- JOINs: They allow you to combine information from various tables, providing a holistic view of your data landscape.

Exploratory Data Analysis (EDA)

"The greatest value of a picture is when it forces us to notice what we never expected to see."

— John W. Tukey

Do you recall when you opened a box of assorted chocolates, uncertain what flavors you'll find inside? You might pick one, take a bite, and examine the texture, flavor, and filling, gradually unraveling the mystery of what's inside. That's the allure of Exploratory Data Analysis (EDA). It's about probing, dissecting, and understanding data, like exploring those chocolates. But instead of taste and texture, you're investigating patterns, anomalies, relationships, and insights your data conceals.

Defining Exploratory Data Analysis

EDA is an initial, essential phase in the data analysis process where you're getting acquainted with your data understanding its structure, quirks, and hidden surprises. It's not about jumping to conclusions or making definitive statements. It's about asking questions, being curious, and allowing the data to guide your exploration.

You may find yourself swamped with numbers, columns, and rows when working with data. It can be overwhelming, like standing at the edge of a vast ocean with no map or compass. EDA is the compass. It helps you navigate, providing direction and clarity as you delve into your analysis.

SQL and EDA: An Unconventional Pairing

SQL, the powerful language of databases, often finds its place in the structured world of querying, fetching, and manipulating data. But can SQL dance to the rhythm of exploration, a process often more chaotic and unstructured? Absolutely!

SQL's structured nature becomes an asset in EDA. It provides a framework and tools that can be wielded with finesse to slice, dice, filter, aggregate, and visualize data.

Think of SQL as a master sculptor's toolkit. In EDA, you're sculpting insights from a block of raw data. And with SQL, you have chisels, hammers, and brushes, each serving a purpose, each helping you shape your sculpture.

SQL's EDA Arsenal

1. Aggregation Functions: AVG, COUNT, MAX, MIN, SUM – these functions help summarize data, providing overviews and insights into distributions.

```
SQL SELECT department, AVG(salary) AS
avg_salary FROM employees GROUP BY
department;
```

This query helps you understand the average salary across different departments, an essential insight for HR analytics.

2. Joins and Subqueries: Combining data from different tables or using nested queries can unravel complex relationships and patterns.

```
SQL SELECT customers.name, COUNT(orders.or-
der_id) AS order_count FROM customers LEFT
JOIN orders ON customers.customer_id =
orders.customer_id GROUP BY customers.name;
```

This query tells you how many orders each customer made, a valuable insight for customer segmentation.

3. Conditional Functions and Case Statements: These help you categorize data based on conditions, aiding in segmentation and pattern recognition.

```
SQL SELECT age, CASE WHEN age < 25 THEN
'Youth' WHEN age < 60 THEN 'Adult' ELSE
'Senior' END AS age_group FROM people;
```

This example categorizes people into age groups, facilitating demographic analysis.

Exploring with SQL: Step-by-Step Examples

Let's embark on a mini-exploration using SQL. Assume you have sales data and want to understand seasonal patterns.

1. **Understand Your Data:** Start with basic queries to know what columns and data types you have. Explore a few rows to get a feel for the data.

2. **Summarize:** Use aggregation functions to find total sales per month.

```sql
SQL SELECT MONTH(sale_date) AS month,
SUM(sale_amount) AS total_sales FROM sales
GROUP BY MONTH(sale_date);
```

3. **Investigate Patterns:** You may notice that sales peak in certain months. Delve deeper by looking at product categories or customer segments.

4. **Visualize:** Sometimes, a graphical representation, though not directly feasible in SQL, can be created using the data fetched with SQL. Visualization often brings patterns and insights to life.

5. **Test Hypotheses:** If you have assumptions or hypotheses, use SQL to test them. For instance, if you think holidays boost sales, you can compare holiday sales versus regular days.

In conjunction with other languages like Python, SQL becomes a versatile and potent tool in EDA. It provides structure to exploration, guiding you through the maze of data, allowing you to ask questions, find answers, and uncover the unexpected.

Hands-On Exercise

 "Tell me, and I forget. Teach me, and I remember. Involve me, and I learn."

— Benjamin Franklin

Allow me to invite you into my workshop for a moment. Here, tools are not just displayed; they are used, honed, and mastered. In this hands-on exercise, we will do just that. Together, we will explore a complex dataset using the advanced SQL queries and EDA techniques you've gathered. It's time to roll up your sleeves and get involved in data discovery.

The Dataset: A World Waiting to be Uncovered

Imagine a vast and diverse dataset filled with information from various domains. It's not just a collection of numbers and strings; it's a representation of the world around us. Whether it's sales data from a multinational company or environmental observations from various locations, a complex dataset is like a treasure trove waiting to be explored.

You can find an example of such a dataset **here**. Download it, load it into your favorite SQL environment, and prepare for an exploration.

Guided Exploration: Uncovering Secrets

This exercise is not a rigid step-by-step manual but a guided exploration. I'll provide directions, propose queries, and explain insights, but the journey is yours. Feel free to wander, ask questions, experiment, and make mistakes. That's how learning happens.

1. Understanding the Terrain: What's in the dataset? What columns? What types of data? What's the range of values? Start with basic queries to get an overview.

```sql
SQL SELECT * FROM dataset LIMIT 5;
```

2. Discovering Patterns: Let's begin to dig deeper. What are the top 10 most common categories? How many unique values are there in a particular column?

```sql
SQL SELECT category, COUNT(*) AS count FROM dataset GROUP BY category ORDER BY count DESC LIMIT 10;
```

3. Investigating Relationships: Are there correlations between certain columns? How do variables interact with each other?

```
SQL  SELECT  region,  AVG(sale_price)  AS
avg_price FROM dataset GROUP BY region;
```

4. Exploring Anomalies: Are there outliers or unusual data points? A query to find the highest values in a column might reveal something unexpected.

```
SQL SELECT product, sale_price FROM dataset
ORDER BY sale_price DESC LIMIT 5;
```

5. Generating Hypotheses: What questions arise from your exploration? Perhaps you notice a seasonal trend or an interesting correlation. Formulate hypotheses and test them using more queries.

 "In God we trust; all others bring data."

— W. Edwards Deming

As we stand on the threshold of another critical phase of our exploration, the words of the renowned statistician resonate with profound wisdom. Data. That's where the truth lies, insights dwell, and our next adventure takes us. After completing the data exploration phase, we've uncovered the landscape mapped the terrains, and now it's time to dive deeper. It's time to explore the realm of statistical analysis using SQL.

Chapter 6
Statistical Analysis

S tatistical analysis isn't merely an analytical process; it's an art that uncovers the hidden melodies within a sea of numbers. It's the bridge between raw data and informed decisions.

Descriptive Statistics: The First Step

We must begin with the fundamentals before we venture into complex statistical modeling. Descriptive statistics provide an overview, summarizing the main aspects of a dataset. They answer questions such as:

- What is the average value?
- What's the range?
- Where is the median, and what does it signify?

Descriptive statistics give us the initial insights, the first glimpse into what the data offers.

L.D. Knowings

Inferential Statistics: Beyond the Obvious

We move to inferential statistics once we understand the data's general characteristics. This is where the real magic happens. Inferential statistics allow us to make predictions, identify relationships, and uncover hidden trends. It's like peering into a crystal ball, but one powered by logic, mathematics, and scientific methods.

Regression Analysis: Finding Connections

One of the powerful tools in statistical analysis is regression. It helps us understand how different variables are interconnected. How does one factor influence another? Can we predict a certain outcome based on known variables? Regression analysis provides these answers, establishing a mathematical relationship that can be both fascinating and incredibly useful.

Hypothesis Testing: The Scientific Approach

In our quest to make sense of data, we often form hypotheses - educated guesses or assumptions about what might happen. Hypothesis testing is validating or refuting these assumptions using statistical methods. It's a rigorous, systematic approach that adds credibility to our analysis.

Time-Series Analysis: Understanding Trends Over Time

Some data unfold over time, revealing temporal patterns. Time-series analysis is the study of these chronological patterns. Whether tracking sales over months, monitoring weather patterns, or analyzing stock market trends, time-series analysis offers invaluable insights.

Your Toolkit: SQL and Beyond

SQL will be our faithful companion throughout this chapter, a powerful tool that allows us to perform these analyses. But it's more than a tool; it's a language that speaks directly to the data, asking questions, seeking answers, and uncovering truths.

Engaging with the Data: Exercises and Tasks

As we delve into these topics, exercises, tasks, and questions will challenge, engage, and deepen your understanding. This isn't a mere theoretical exercise; it's a hands-on, practical exploration.

 "The first step toward change is awareness. The second step is acceptance."

— Nathaniel Branden

Descriptive Statistics: The Unsung Hero of Data Analysis with SQL

> *"Statistics are like bikinis. What they reveal is suggestive, but what they conceal is vital."*

— Aaron Levenstein

You might find this quote cheekily appropriate if you've been with me through this SQL adventure. You're about to dip your toes into the pool of descriptive statistics, and I promise, it won't be as cold as you think.

Why Descriptive Statistics Matter in Data Analysis

Imagine gazing at a colossal spreadsheet with numbers, columns, and rows. Should you accept it, your mission is to extract something meaningful from this sea of numbers. A daunting task, right? Here's where descriptive statistics saunter, like the unsung hero in an action-packed thriller.

You see, descriptive statistics are the bread and butter of data analytics. They offer a snapshot of the data, summarizing its main aspects in a digestible way. Think of it like this: If SQL is the camera, descriptive statistics are the carefully chosen Instagram filters that highlight the best features while concealing the unnecessary clutter. They form the initial layer of understanding before you dive into deeper analyses.

Chapter 6

Essential SQL Functions for Descriptive Statistics

Now, let's roll up those sleeves and get our hands dirty with some SQL magic. The language comes loaded with built-in functions tailored for descriptive statistics. Ah, you've got to love SQL for its versatility.

- COUNT(): How many rows of data do you have? COUNT() answers that.
- AVG(): Want to know the average value of a certain column? Enter AVG().
- MIN() and MAX(): The lowest and highest values? MIN() and MAX() have you covered.
- SUM(): Need the total sum of a numeric column? SUM() is your go-to.
- GROUP BY: Want to break down these statistics by categories? GROUP BY is your trusty sidekick.

Calculating Measures of Central Tendency with SQL

Okay, enough talk. Let's get down to some SQL action. Measures of central tendency—mean, median, and mode—are essential to understanding the "center" of your data distribution.

- Mean: The average is calculated using SQL's AVG() function. Let's say you want to find the average salary in a company. Your SQL query would look something like this: SELECT AVG(salary) FROM employees;

- Median: Ah, the middle child of central tendency. Unfortunately, SQL doesn't have a built-in function for median, but don't fret. You can find it by sorting the data and selecting the middle value. It's a little cumbersome, I know, but it's doable.
- Mode: This value appears most often in your dataset. You can calculate the mode by grouping the data and using COUNT() to find the most frequent value.

Understanding Measures of Dispersion

If measures of central tendency are the quarterbacks on the data analysis football team, then measures of dispersion are the unsung linemen protecting them. These include range, variance, and standard deviation.

- Range: The simplest measure, calculated as MAX() - MIN(). If you have a column of ages, you could find the range with: SELECT MAX(age) - MIN(age) FROM people;
- Variance: This measures how spread out the values are from the mean. SQL lacks a built-in function for variance, but you can calculate it using AVG() and some arithmetic.

- Standard Deviation: This is simply the square root of the variance. Again, some SQL gymnastics is required, but it's worth the stretch.

Practical Examples: SQL Queries Illustrating These Concepts

Let's make this real with some SQL queries. Suppose we have a table called student_scores with columns student_id and score.

- To find the mean score:

```
SELECT AVG(score) FROM student_scores;
```

- To find the mode score:

```
SQL SELECT score, COUNT(*) FROM student_s-
cores GROUP BY score ORDER BY COUNT(*) DESC
LIMIT 1;
```

- To find the range of scores:

```
SQL SELECT MAX(score) - MIN(score) FROM
student_scores;
```

- To find the variance:

```
SQL SELECT AVG((score - (SELECT AVG(score)
FROM student_scores)) * (score - (SELECT
AVG(score) FROM student_scores))) FROM
student_scores;
```

Hypothesis Testing: Cracking the Code of Uncertainty with SQL

> *"Walking on water and developing software from a specification are easy if both are frozen."*
>
> — Edward V. Berard

Ah, the paradox of uncertainty. We humans have a love-hate relationship with it, don't we? On one hand, it keeps life interesting; conversely, the pesky variable always seems to throw off our calculations. But what if I told you SQL could help you make friends with uncertainty, at least regarding data analytics? Intriguing, right? Let's dive in.

The Art and Science of Hypothesis Testing

Hypothesis testing is like solving a mystery. Based on a sample dataset, you start with a question or an assumption about a population. This assumption is what we call the "null hypothesis." It's your default position, the claim you want to test. Then there's the "alternative hypothesis," which you want to prove instead.

The process of hypothesis testing is akin to a court trial. The null hypothesis assumes "innocence" (or no effect), and you, the data analyst, play the role of the prosecutor trying to prove "guilt" (or an effect). You collect evidence (data), analyze it, and then determine if it's strong enough to reject the null hypothesis in favor of the alternative one.

Now, you might be wondering how SQL fits into this picture. SQL is your investigative toolkit, offering a range of functions and methods to conduct these tests efficiently. It lets you filter data, perform calculations, and even automate repetitive tasks to pursue evidential truth.

SQL Tools for Your Hypothesis Testing

Let's say you're working on a project where you need to prove that a new marketing strategy has increased sales. Your null hypothesis could be: "The new marketing strategy does not affect sales." And your alternative view? "The new marketing strategy has increased sales."

In SQL, you can use various aggregate functions to calculate averages, sums, and even variances of sales before and after implementing the new strategy. You can also use conditional clauses to isolate specific data sets, making your hypothesis test more targeted.

- COUNT() and AVG(): For counting the number of sales and calculating the average value.
- SUM() and STDDEV(): To find the sum of all sales and calculate the standard deviation.
- CASE WHEN: Categorizing data based on conditions is practical when comparing different groups.

Practical SQL Queries for Hypothesis Testing

You're all about practicality, so let's look at some SQL examples to understand better how hypothesis testing is done in real life.

Imagine we have a database table named SalesData with columns SaleID, Date, and Amount. To analyze the impact of the new marketing strategy implemented on '2021-01-01', we can write SQL queries like these:

1. To calculate the average sales before the strategy:

```
SQL SELECT AVG(Amount) FROM SalesData WHERE
Date < '2021-01-01';
```

2. To calculate the average sales after the strategy:

```
SQL SELECT AVG(Amount) FROM SalesData WHERE
Date >= '2021-01-01';
```

3. To find the standard deviation of sales after the strategy:

```sql
SQL SELECT STDDEV(Amount) FROM SalesData
WHERE Date >= '2021-01-01';
```

These SQL queries will provide you with the raw numbers. You can then use statistical methods to determine if the differences are statistically significant, helping you make informed decisions.

A Quick Self-Check Exercise

To ensure you've got the hang of it, try writing an SQL query to find the sum of all sales after implementing the new strategy. Don't peek; try doing it on your own first.

Done? Great! The query would look something like this:

```sql
SQL SELECT SUM(Amount) FROM SalesData WHERE
Date >= '2021-01-01';
```

Analyzing Relationships: Making Sense of Correlation, ANOVA, and Regression Analysis with SQL

 "Numbers have an important story to tell. They rely on you to give them a voice."

— Stephen Few

Ah, the elusive relationships between variables, the Rosetta Stones of data analytics! Why does one variable change when another does? What patterns lurk in the mire of numbers and codes? Look, I get it. This is the stuff that keeps you up at night. As someone who's spent much of my life in the IT trenches, I understand the allure and the challenge.

The What and Why of Correlation

Let's start with correlation. It's all about the relationship. How two variables change about each other. If one goes up, does the other one follow suit? Or do they move in opposite directions? In the SQL universe, correlation is not just a concept; it's an actionable metric. With the CORR() function, SQL provides a Pearson correlation coefficient between -1 and 1 that tells how these variables interact. A positive value signifies a positive correlation, a negative value suggests an inverse correlation and a value close to zero. Well, those variables are as indifferent to each other as cats are to the concept of your personal space.

ANOVA: The Variance Decoder

Moving on, let's talk about ANOVA or Analysis of Variance. When you have multiple groups, and you're itching to find out if they're all cut from the same cloth, ANOVA is your go-to. For instance, does the average salary differ significantly between different departments in a company? SQL takes the sweat out of this complex analysis. Functions like AVG, COUNT, and SUM are your comrades-in-arms here, allowing you to calculate means, variances, and the F-statistic that ultimately helps you understand everything.

Regression Analysis: The Future Predictor

Ah, regression analysis, the soothsayer of the data analytics world. It goes beyond just identifying relationships between variables to making actual predictions. Imagine knowing the future sales of a product based on advertising spend. SQL offers a palette of functions for regression analysis, including but not limited to LINEST and LOGEST. This is where SQL's real muscle shows, helping you fit a line (or a curve for the adventurous) through your data points.

SQL in Action: Practical Examples

Now for the part you've been waiting for: how to do this in SQL. Ready to roll up those sleeves?

For correlation, a simple SQL query can look like this:

```
SQL SELECT CORR(column1, column2) FROM table_name;
```

For ANOVA, while SQL doesn't have a direct ANOVA function, you can use a combination of AVG and COUNT to calculate group means and variances. Here's a very simplified example using the AVG function:

```
SQL SELECT department, AVG(salary) FROM employees GROUP BY department;
```

In the case of regression, SQL offers a bit more in the way of direct functionality:

```
SQL SELECT LINEST(y_values, x_values) FROM table_name;
```

The So-What Factor: Predictive Analytics and Decision-Making

So, why should anyone care about correlation, ANOVA, or regression analysis? Because these aren't just statistical gymnastics. They're the backbone of predictive analytics. These methods enable you to make data-based forecasts, not just gut feelings. And in today's data-driven world, that's pure gold. Whether you're deciding on a new product launch, a shift in strategy, or identifying potential market trends, these analyses offer actionable insights.

Hands-On Exercise: Roll Up Your Sleeves for Statistical Analysis

 "If you can't explain it simply, you don't understand it well enough."

— Albert Einstein

Ah, the aroma of freshly brewed coffee, the faint tapping of keys, and a dataset spread out before you like a treasure map. There's something magnetic about the prospect of diving into a new challenge, isn't there?

Your Laboratory: The Sample Dataset

So here's the deal: I'm providing you with a sample dataset. Think of it as a sandbox where you'll play with all the shiny SQL toys we've discussed. You're not just pressing your nose to the glass of the toy store; you're inside, trying out each one.

Here's a simplified dataset of a hypothetical online retail store:

```SQL
CREATE TABLE online_sales ( sale_id INT PRIMARY KEY, product_name VARCHAR(50), quantity_sold INT, price_each FLOAT, sale_date DATE );

INSERT INTO online_sales VALUES (1, 'Smartphone', 2, 299.99, '2022-01-01'), (2, 'Laptop', 1, 899.99, '2022-01-02'), (3, 'Headphones', 5, 49.99, '2022-01-05'), ... ;
```

Statistician Mode: Descriptive Statistics

First, calculate some descriptive statistics to get a feel for our data. How about calculating the average price of all the products sold?

```
SQL SELECT AVG(price_each) FROM online_sales;
```

And while we're at it, let's find the most and least frequently sold products:

```
SQL SELECT product_name, COUNT(*) as frequency FROM online_sales GROUP BY product_name ORDER BY frequency DESC;
```

The Bond Between Variables: Correlation Analysis

Now, let's heat things a bit. Ever wonder how the quantity of products sold correlates with their prices? Is cheaper always better, or do some high-ticket items defy the odds? You're about to find out. Use SQL's CORR() function to determine the correlation between quantity_sold and price_each.

```
SQL SELECT CORR(quantity_sold, price_each)
FROM online_sales;
```

The Crystal Ball: Regression Analysis

Finally, let's move on to regression analysis. Imagine being able to predict future sales based on the product's price. Yep, it's almost like peering into a crystal ball. And SQL is your friendly neighborhood wizard here.

```
SQL -- For demonstration, actual SQL syntax
for regression may differ based on your SQL
version SELECT LINEST(price_each, quanti-
ty_sold) FROM online_sales;
```

Interpret the Symphony of Numbers

Okay, the numbers are in. The most rewarding part comes from interpreting what these numbers whisper to us. If the correlation value is close to 1 or -1, you've got a strong relationship between price and quantity sold. What could that mean for your pricing strategy?

For regression analysis, what did the output tell you? Can you predict how altering the price might affect the number of units sold?

The Bigger Picture: Implications and Insights

It's not just about number-crunching. It's about what these numbers tell us—about consumer behavior, potential roadblocks, and golden opportunities. Maybe you discovered that the cheaper products aren't always the most frequently sold, defying conventional wisdom. Or perhaps the regression analysis hinted at an optimal price point that maximizes profit and sales volume.

> *"The Function of Good Software is to Make the Complex Appear Simple"*
>
> — Grady Booch

Ah, yes. Complexity, that old nemesis we've been gently nudging aside, step-by-step, SQL query by SQL query. You've done well so far, wrestling with data preparation, integration, and statistical analysis. You've been the unsung hero, tidying up messy data and pulling insights from seemingly mundane tables. Now, how about we make it all visually stunning? Brace yourself because our next focus will make your data speak and sing — Data Visualization.

A Picture Worth a Thousand Queries

See, no matter how eloquent your SQL queries are or how insightful your statistical analyses may be, if you can't communicate these findings clearly, they're like a masterpiece painting locked away in a vault. Visualization is your key to bringing this masterpiece into the light, to let it be admired, scrutinized, and most importantly, understood. And let's face it, a well-crafted chart can often convey what a 100-line SQL query can't.

The 'Why' Behind the 'What'

But hold on a second. Why is data visualization so crucial? Human brains are wired to process visual information far more efficiently than text. That means the quicker you translate your SQL findings into visual elements, the faster your audience can understand and act on them. Plus, for those in industries like finance, marketing, or healthcare, mastering data visualization isn't just a fancy skill to add to your resume; it's a necessity.

Challenges, Meet Solutions

You may have already encountered hurdles when you tried to visualize complex datasets. Maybe you've dealt with unmanageable sizes, inconsistent formatting, or didn't know which chart type would best represent your data. I get it; it can be daunting. But fret not because we're going to tackle these challenges head-on. You'll soon be creating visualizations that look good and make sense.

The Toolbox

You'll be thrilled that SQL offers some elementary data visualization capabilities. Of course, other specialized tools like Tableau, Power BI, and R's ggplot2 are worth exploring later. But for now, knowing how to generate a simple bar chart or pie chart directly from your SQL interface can be remarkably empowering.

Here's a quick task list to whet your appetite for what's coming:

- Task 1: Create a bar chart to represent monthly sales data.
- Task 2: Use a line graph to track price fluctuations.
- Task 3: Generate a pie chart to show the distribution of customer demographics.

Taking it a Step Further

But we won't just stop at creating pretty pictures. You'll learn how to make your visualizations interactive, use color psychology to make your points more compelling, and craft narratives around your visual data. You'll essentially become not just a data analyst but a data storyteller.

Chapter 6

Got Your Curiosity Piqued?

What kind of data are you most excited to visualize? Are there patterns you've noticed that you're itching to represent visually? How could a compelling visualization change how your organization sees a particular issue?

Your Stage Awaits

In the upcoming chapter, we'll get our hands dirty with the nuts and bolts of data visualization. From choosing the right charts to understanding the principles of visual hierarchy, we'll cover it all. You'll go from being the person who knows how to extract data to the one who knows how to make it dance — ah, sorry, I mean, make it 'come alive.'

So, as you flip the page, prepare to step into a world where numbers transform into shapes and colors, where data points morph into visual stories, and where your SQL skills get their well-deserved spotlight. Trust me, it's a game-changer.

Chapter 7
The Power of Seeing: Data Visualization Unleashed

"Sometimes the questions are complicated, and the answers are simple. But how do you reveal those simple answers hidden in complex data?" Dr. Seuss never faced the enigma of SQL databases, but his wisdom is surprisingly relevant. If you've ever found yourself staring at rows upon rows of SQL data, you've likely wondered how to translate those numbers into something more human. That's where data visualization comes in. It's the symbolic artist's palette of data analytics.

Why Visualization Matters

Let's get real. You're not just crunching numbers for the sake of it. Whether you're a data analyst in a high-stakes financial firm or a researcher sifting through health statistics, you're driven by questions that matter. Questions that could mean the difference between a thriving business and a floundering one, between effective healthcare policy and a systemic failure. But numbers alone won't give you the whole picture.

Here's the thing about data—its true power lies in the patterns it forms, the trends it reveals, the stories it tells. And what's the best way to discern patterns or trends? By seeing them. Literally. The human brain processes images 60,000 times faster than text. Just think about that for a second. The sheer speed at which you can absorb and interpret information when it's presented visually is astounding. The right graph or chart can speak volumes, revealing the very soul of the data.

Your Tools of the Trade: Tableau and Power BI

Now that we've established why visualization is your secret weapon let's talk tools. We've got some heavy hitters in the industry: Tableau and Power BI. These aren't just fancy software; they are your allies on the battlefield of data analysis. Tableau's intuitive interface allows you to drag and drop your way to stunning visuals. At the same time, Power BI offers seamless integration with other Microsoft products, making it a go-to choice for many organizations.

Why these tools, you ask? Because they are designed to be used by people like you. You don't need a Ph.D. in computer science to operate them. You can transform an incomprehensible sea of numbers into a compelling story with a few clicks. And remember, data visualization isn't just about making things look pretty. It's about clarity, precision, and

—yes, I'll say it—revelation. When your boss asks why sales dipped last quarter, you won't just hand over a spreadsheet; you'll present a narrative illustrated with striking visuals that get straight to the point.

From SQL to Visual Storytelling

Let's connect the dots back to SQL. You've learned to query databases, manipulate data, and perform advanced analyses. That's fantastic. You're a wizard with words like "SELECT," "FROM," and "WHERE." But now it's time to transform that raw SQL data into something more digestible.

How? Easy. Both Tableau and Power BI allow you to import SQL data directly. There is no need for complicated conversions or clunky workarounds. Once your data is imported, you're free to play. Do you want to compare monthly revenue for the past year? There's a chart for that. Are you interested in geographical trends? Plot it on a map. The possibilities are as endless as your curiosity.

The Practicalities: A Hands-On Guide

You've got your SQL data, and you've chosen your tool. What next? It's not magic, but it's close. Import your SQL tables into Tableau or Power BI. Trust me, it's straightforward. Once you're in, it's like being a kid in a candy store, except the candy is data, and you're not a kid; you're a professional with a mission. Drag and drop different variables, apply filters, and select the type of chart or graph that tells your story best. Before you know it, you'll have visuals that don't just represent your data—they embody it.

The Unseen Canvas: Visualizing SQL Data as Art

"Art washes away from the soul the dust of everyday life," said Pablo Picasso. In the seemingly monotonous world of SQL data, where tables and queries often blur into a routine, consider visualization as your art form—cleansing your analytic soul. No, you don't have to be Picasso or Monet, but you do need to understand that raw SQL data is your canvas, and your visualization tools are your paintbrushes.

The Silent Language of Data

I get it. You've spent hours, maybe even days, sifting through tables, writing SQL queries, and wrestling with inner and outer joins. And after all that, you're left with more tables. It's like staring at sheet music without hearing the melody. You understand the notes, but you're not feeling the music. Visualization lets you hear that silent language, transforming notes into symphonies and numbers into narratives.

From Spreadsheets to Storyboards

Visuals are the storyboard to your data's screenplay. They give form to your data's voice, adding color to its stories, just as a storyboard does for a script. Imagine you're a researcher in healthcare; you're not just plotting graphs, you're mapping out patterns that could influence medical policies. If you're in finance, your pie charts and histograms could be the difference between a winning investment and a fiscal faux pas.

The Palette: Tableau and Power BI

Enough with the artistic metaphors; let's talk brass tacks. Or should I say, brushes and palettes? The tools you choose can make or break your visualization. Your go-to's are Tableau and Power BI. Why? Because they get you and the struggles you've had with SQL data. They know you need something user-friendly, a gentle learning curve, and something robust enough to satisfy your inner data artist. Tableau is like that art teacher who encourages free expression—drag, drop, and discover. Power BI, on the other hand, is the meticulous instructor, making sure your lines are straight, and your colors are within the borders, especially if you're already cozied up with Microsoft's ecosystem.

The First Brush Stroke: Importing SQL Data

How do you start? You've got your SQL database brimming with tables just begging to be understood. First, you import them into either Tableau or Power BI. This isn't rocket science or abstract art; it's as straightforward as choosing a file to upload. And guess what? Both platforms support direct SQL data imports. So, no more juggling between applications or tedious data conversions. It is just a smooth transition from raw data to ready-to-paint canvas.

The Art of Interactive Storytelling

Interactive storytelling is an art, and you're the artist. Think about it. When you're in the zone, dragging and dropping variables, applying filters, and toggling between different types of charts, you're not just analyzing data; you're telling a story. Your audience? Anyone from your team leader to the CEO, each waiting to see your data come to life. The interactive nature of tools like Tableau and Power BI allows your audience to dive into the story, explore its layers, and even change its outcome. Now, that's powerful storytelling.

Your Masterpiece: The Final Visualization

So, you've got your imported SQL data, messed around with variables, applied filters, and selected the most fitting charts. Now comes the moment of truth: the final visualization. This isn't just any chart or graph; it culminates your analytical thought process. It's your masterpiece, ready to be showcased in the gallery of business decisions. Whether revealing consumer behavior patterns or providing actionable insights for your next marketing strategy, your visualization is more than a display; it's a declaration. It says, "Here's what the data tells us, clear as day, plain as paint."

A Canvas in Numbers: The Imperative of Data Visualization

"Data is the new oil," they say. But just like crude oil needs refining to fuel your car, raw data needs visualization to fuel your decision-making. Piles of raw data can seem as chaotic as a Pollock painting. But when organized, sifted through, and presented visually, it becomes a Mona Lisa—every data point adds to a meaningful whole.

Gazing into the Crystal Ball: The Definition and Vitality of Data Visualization

Let's clarify: Data visualization isn't some highfalutin term to intimidate the SQL newbie or the aspiring data scientist. It's not a luxury; it's a necessity. Data visualization is your compass in the wilderness of zeros and ones. It's taking your meticulously queried data and translating it into something digestible—a bar graph, a pie chart, a heat map. Anything that makes complex data understandable at a glance.

Why is this important? Because we're human. Despite our love for numbers and algorithms, our brains are wired for visual input. Studies have shown that the human brain processes images 60,000 times faster than text. No wonder a pie chart communicates more in a second than a spreadsheet does in fifteen minutes! Visualization reveals patterns, unveils trends and highlights correlations that a simple table of numbers can't. Visualizing data is like revealing the cards you didn't know you were holding.

The Oracle's Tools: Discovering the Undiscovered in Data

Let's say you're a business analyst in a healthcare startup or a data scientist in a Wall Street firm. You've got this SQL database teeming with data, each table and row pregnant with potential insights. What's the next step? It would be best if you saw the story hiding in the statistics. You've got tools for this—Tableau, Power BI, heck, even good old Excel charts can work wonders if used creatively. These are not mere software; think of them as your oracles. They reveal what the plain numbers keep veiled: the trends, the patterns, and anomalies.

The Unveiling: Discovery through Visualization

Do you remember those "Magic Eye" books from the '90s, where a pattern of shapes and colors would reveal a 3D image if you stared at it long enough? That's what happens when you visualize data. You're looking at the same numbers and variables, but now they reveal something entirely new—something crucial. Trends in patient recovery rates, quarterly profits, and customer engagement spikes are the "3D images" your data has been hiding.

Reality Checks: Trusting Your Visualizations

Alright, a word of caution here. Visualization is powerful, but it's not foolproof. Remember the old saying, "Lies, damned lies, and statistics"? Well, it can apply to data visualization, too. A pie chart can easily misrepresent data if the slices aren't proportioned correctly. A line graph can give an exaggerated sense of trend if the axes are not scaled right. So, while the tools are great, they're not a substitute for your discernment. Make sure to double-check, triple-check if you have to.

Your Toolkit: Exercises for Effective Visualization

Let's not just talk; let's do. Open your SQL software and pull up a dataset—any dataset you are working on or something from your past projects. Now, import this into Tableau or Power BI.

Task List for Visualization:

- Import your SQL data into Tableau or Power BI
- Create a basic bar chart using one of the variables

- Now, try a more complex visualization like a heat map or a bubble chart

These exercises are your playground. Make mistakes, find flaws, and then correct them. That's how you'll learn. That's how you'll get better.

The Invisible Hand of SQL in Crafting Visual Masterpieces

> *"Good data visualization is like a Zen garden—each rock, each shrub, each rake line is placed with purpose and intent. SQL is the gardener."*

SQL: The Sculptor of Raw Data

SQL often wears the crown for data extraction and manipulation in data analytics. But let's not forget another role it plays, a subtler yet powerful one—that of a sculptor, chipping away the excess to reveal the beauty within. No, SQL doesn't create visualizations but prepares the canvas, selects the palette, and even sketches the outlines. When wrestling with complex SQL queries, remember this: You're not just manipulating data; you're setting the stage for it to come alive visually.

Let's pause for a moment for those of you knee-deep in lines of SQL code, fretting over complex JOIN operations or intricate subqueries. Why are you doing this? To create reports? To analyze trends? Sure, but let's step back even further. You're doing this to understand, to grasp the hidden narratives within heaps of data. SQL queries serve as your eyes, extracting precisely what you need to see. But often, for a more profound understanding, your data requires a voice—and that voice is visualization.

The Marriage of SQL and Data Visualization

SQL and data visualization are like the Lennon and McCartney of data analytics. They're brilliant on their own but unstoppable together. Let's imagine you're a data analyst in the healthcare sector. Your SQL queries can identify exact patient demographics, diagnosis rates, and treatment successes. But can these tables of data show you patterns at a glance? Unlikely. That's where visualization tools come in. They transform SQL outputs into insightful visual narratives—bar graphs, pie charts, heat maps—you name it.

Crafting the Perfect Data Set with SQL

SQL is where the initial magic happens before you even open your Tableau or Power BI. SQL allows you to filter out the noise. It lets you curate the data points that are most relevant for your visualization. Think about it. By selecting only the fields you need, aggregating data, or creating calculated fields, you are shaping the raw data into something more malleable and meaningful. It's like cooking; you start with raw ingredients, but how you cut, season, and combine them makes all the difference in the final dish.

SQL Functions Tailor-Made for Visualization

SQL isn't just a one-trick pony. It's equipped with functions that are tailor-made for preparing data for visualization. Functions like GROUP BY allow you to categorize data perfectly for pie charts or bar graphs. Temporal functions can arrange your data chronologically, setting it up for a compelling timeline or a line graph. Statistical functions can provide summaries of data sets, which can then be visually represented as box plots or histograms.

What's Your Task? Please test it Out!

Let's get those hands dirty. If you've got a SQL interface up and running, here's what you should try.

Task List:

- Run an SQL query to extract data, but use the GROUP BY function to categorize it.
- Use an ORDER BY clause to sort data chronologically.
- Experiment with SQL statistical functions like AVG or SUM to get summary data.

After you've done this, take this refined data into a visualization tool of your choice. Observe how much easier and more insightful the process becomes when your data is prepped and primed.

The Art of Visualizing SQL Data: Choosing Between Tableau and Power BI

 "In the age of information overload, the most valuable currency is clarity. Data visualization tools are the translators, and SQL is the dialect they understand."

Tableau: The Maestro of Interactive Dashboards

You've nailed down your SQL queries and extracted that invaluable data, and now what? Staring at rows upon rows of data isn't exactly a feast for the eyes or fodder for the mind. Enter Tableau, the maestro of turning your SQL outputs into interactive, insightful dashboards. Oh, I get it; you've heard of it. But have you used it?

Tableau is a powerhouse when it comes to data visualization. It allows you to create everything from simple bar charts to intricate geographical heat maps. And let's not forget its robust compatibility with SQL databases. You can connect directly to your SQL server, import the data, and you're ready to unleash your inner artist. Its key features include real-time data analytics, a user-friendly drag-and-drop interface, and seamless sharing of your dashboards with your team. And for those of you who are a tad more advanced, Tableau also offers scripting capabilities. You can almost hear the data singing, can't you?

Task List for Experimenting with Tableau:

- Connect Tableau to your SQL database.
- Import a simple dataset.
- Try creating a bar chart and a heat map.
- Share your dashboard with a colleague and ask for feedback.

Power BI: Microsoft's Gift to Data Lovers

Power BI might be more up your alley if you're a Microsoft enthusiast. This is another heavy hitter in the data visualization world. Think of Power BI as the multi-tool Swiss knife for data analysts. Need to create interactive reports? Check. Want real-time dashboard updates? Double-check. Need seamless SQL integration? Oh, you bet that's a check.

Power BI lets you connect directly to SQL databases, offering native SQL query support. Moreover, it integrates well with other Microsoft Office products—no surprise. With Power BI, you get quick measures, forecasting, and grouping features. It also lets you publish your reports to the Power BI service, where you can create dashboards and share them with your organization.

Chapter 7

Task List for Experimenting with Power BI:

- Connect Power BI to your SQL database.
- Import your data into the Power BI Desktop.
- Create at least two different types of visualizations.
- Publish your report to the Power BI service.

Tableau vs. Power BI: The Showdown

Alright, so which one is the MVP in the world of data visualization tools? Well, it depends on what you're looking for. Tableau shines when it comes to user experience. Its interface is intuitive, and the drag-and-drop functionality makes you feel like a kid with a new set of LEGO blocks. However, if budget constraints make you wince, be aware that Tableau can be pricier.

On the other hand, Power BI excels in real-time analytics and data-sharing capabilities. If your organization is already committed to Microsoft products, Power BI will slide into your workflow like the last piece of a jigsaw puzzle. But it's worth mentioning that Power BI's interface can feel a bit clunky, especially if you're used to the fluidity of Tableau.

So, which one should you choose? Do you value a more user-friendly interface and broader visual capabilities? Tableau is your pick. But if you prioritize real-time analytics and are already living in the Microsoft ecosystem, Power BI is your go-to.

Table for Quick Reference:

Feature	Tableau	Power BI
Ease of Use	Intuitive	Slightly Clunky
Real-Time Analytics	Good	Excellent
Pricing	Premium	More Affordable
SQL Support	Robust	Native Support
Dashboard Sharing	Seamless	Integrated with Microsoft Office

Making Sense of the Mess: SQL Techniques for Data Preparation

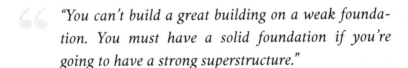

"You can't build a great building on a weak foundation. You must have a solid foundation if you're going to have a strong superstructure."

— Gordon B. Hinckley

Why the Fuss About Data Preparation?

Don't let anyone tell you otherwise—data preparation is the unsung hero of data analytics. That behind-the-scenes stagehand ensures the spotlight hits you at just the right angle. The curtain rises, and voila, your data is ready for its grand performance in the form of compelling visuals. But, ah, we're getting ahead of ourselves. Let's rein it in. So, why is data preparation so crucial?

Well, think of data as raw material. It arrives in heaps, disorganized, and chaotic. And while SQL is your robust tool for querying this data, you can't extract valuable insights from it unless it's cleaned, structured, and well-oiled. The cleaner and more organized your data, the more accurate your visualizations will be more precise and insightful. The role of data preparation is to make this chaos understandable, to transform it into a language that your data visualization tools can understand effortlessly.

Practical Exercise:

- Run an SQL query on your database to check for missing or inconsistent data.
- Take note of what needs to be cleaned or transformed.

SQL: Your Swiss Knife for Data Preparation

I've got to level with you. SQL is not just for querying data. Its utility in data preparation is often overlooked, but let me assure you, it's a game-changer. Remember the mess we talked about? SQL is your Marie Kondo for data; it tidies up the place. It can handle missing data, manage outliers, and even transform data types to make them more suitable for analysis.

Let's say you're dealing with a column of dates in different formats. SQL functions like CAST and CONVERT can unify them. What about those pesky NULL values messing up your averages? The COALESCE function can replace them with zeros or any default value. Need to deal with outliers? SQL provides a host of statistical functions to identify and neutralize them. And let's not even get started on how SQL can aggregate and summarize your data, making it ripe for high-level visualizations. In essence, SQL prepares your data so it's already a well-rehearsed act when it reaches the visualization stage.

Practical Exercise:

- Clean your data with SQL functions like CAST, CONVERT, and COALESCE.
- Apply aggregation functions like SUM, AVG, and COUNT to prepare summary data.

SQL Functions for Data Preparation

Now for the good stuff—SQL functions you can use to prep your data. Don't just skim through this section; these functions are your bread and butter for data preparation. Functions like TRIM to remove extra spaces, UPPER and LOWER to standardize text cases, and ROUND to handle decimal numbers are just the tip of the iceberg. When dealing with time-series data, functions like DATEPART and DATEDIFF can split hairs into temporal data. And that's not all. SQL offers a whole suite of mathematical functions that help you engineer new features from existing ones, which can be particularly helpful when visualizing complex relationships in the data.

Think of these functions as your prep kitchen gadgets; they make the process efficient, letting you focus on the flavor (i.e., insights) rather than getting stuck peeling and chopping.

Practical Exercise:

- Use text functions like TRIM, UPPER, and LOWER to clean text data.
- Apply time functions like DATEPART and DATEDIFF on your time-series data.

"Show, Don't Tell" in Data: Mastering Interactive Visualizations

 "Data is just like crude. It's valuable, but if unrefined, it cannot really be used."

— Clive Humby

A Palette of Visual Possibilities

We've spent much time wrestling with SQL queries, manipulating data sets, and diving deep into analytics. Now, we're stepping into the vibrant world of data visualization. Ah, yes, the grand reveal where numbers transform into bar graphs, pie charts, and heat maps. Why? Because humans are visually wired beings. We process visuals 60,000 times faster than text. In a way, visualizations are the art galleries of the data world, and you, my friend, are the curator.

But before you can paint the Sistine Chapel, you need to master your strokes. I want you to understand how to create pretty visuals and tell compelling stories. How do you decide between a pie chart and a scatter plot? When should you opt for a heat map? Let's untangle these choices.

Thought-Provoking Question: What was the last visualization you saw that stuck with you? Why did it leave an impression? Reflect on this as you explore the different types of visualizations.

Chapter 7

SQL: The Underpinning of Your Art

Let's not forget the unsung hero—SQL. Your meticulously prepared SQL datasets are the canvases upon which your visual masterpieces will unfold. Whether you're using Tableau or Power BI, the first step is ensuring your data is in tip-top shape. Missing values? Eliminate or replace them. Inconsistent formats? Standardize them. It might seem tedious, but every bit of effort you put into data prep with SQL pays off tenfold in the clarity and accuracy of your visualizations.

Practical Exercise: Take a small dataset and run an SQL query to identify null or missing values. Replace them using the SQL COALESCE function.

Your Tools: Tableau or Power BI?

Both Tableau and Power BI are behemoths in the world of data visualization, each with its own set of strengths and quirks. Tableau is a wiz at creating intricate, detailed visualizations. It's like the painter who delicately crafts each brushstroke. Power BI, on the other hand, is incredibly user-friendly and integrates seamlessly with other Microsoft Office products. Think of it as the digital artist who leverages technology for ease and efficiency.

L.D. Knowings

The Workshop: Hands-On Exercise in Interactive Visualization

Alright, enough talk. Let's get our hands dirty. I've prepared an exercise to guide you through creating charts and dashboards using Tableau or Power BI. You will interpret the visualizations and draw conclusions based on the data you see. There's nothing like applying what you've learned to solidify your understanding.

Dataset: A sample dataset containing sales data for a retail company. Fields include ProductID, ProductName, Category, QuantitySold, TotalSalesValue, and DateOfSale.

Steps:

1. Data Preparation: Import the dataset into your SQL environment and run queries to clean the data. Check for missing values and inconsistencies.
2. Export to Visualization Tool: Once your data is cleaned and prepped, export it to either Tableau or Power BI.
3. Creating Charts: Start with a simple bar chart to represent sales by product categories. Gradually move to more complex visualizations like pie charts for sales distribution and heat maps for sales over time.
4. Dashboard Creation: Combine these visual elements into a comprehensive dashboard.

5. Interpretation and Conclusions: Spend time analyzing your dashboard. What story does it tell? Are there any surprising trends or outliers?

Practical Exercise:

- Import the sample dataset into your SQL tool.
- Run SQL queries for data cleaning and preparation.
- Export the dataset to your chosen visualization tool and follow the steps to create charts and dashboards.

Chapter 8
The Future Is Big Data, and SQL Is the Key to Unlocking It

 "Numbers have an important story to tell. They rely on you to give them a clear and convincing voice."

— Stephen Few

Data: Your Raw Material, Your Goldmine

We've been knee-deep in SQL queries, navigated the labyrinth of data analytics, and even played the role of artists in data visualization. So, what's next? Ah, we're staring at the behemoth: Big Data. For many, Big Data conjures images of an insurmountable mountain of information. But let's reframe that perspective—think of it as an abundant goldmine, teeming with potential insights that could change how you work, your company's operations, or even how the world understands a particular issue.

Thought-Provoking Question: Have you ever felt over-whelmed by the sheer volume of data you have to deal with? How did you manage it, and what could you have done differently?

SQL, The Indispensable Guide to Your Big Data Expedition

It's tempting to think of Big Data as a separate entity that requires specialized, arcane knowledge. But here's the kicker: the SQL skills you've been honing are your ticket into this complex world. That's right, SQL isn't just for your run-of-the-mill databases; it's a versatile tool that can tackle Big Data's volume, velocity, and variety. Picture SQL as your all-terrain vehicle, robust and agile, capable of navigating the hilly terrains and deep valleys of Big Data.

SQL: Not Just a Query Language, but Your Big Data Power Tool

They say that everything looks like a nail when you have a hammer. SQL is more than just a hammer in your data analysis toolkit; it's a Swiss Army knife. For Big Data, SQL can perform data aggregations on a massive scale, run intricate joins across tables with billions of rows, and even execute text-based searches. No, you don't need to learn another programming language or tool; mastering SQL gives you the keys to unlock the treasures hidden in Big Data.

Chapter 8

Surprising Statistic: Did you know that around 90% of the data in the world today has been created in the last two years alone? That's the magnitude of Big Data we're talking about.

Segue: Transition to the Topic of SQL and Big Data, to be Covered in the Next Chapter

We've scratched the surface of what SQL can do with Big Data. But hold onto your hats because the next chapter dives into the nitty-gritty details, from data lakes and data warehouses to real-time analytics. We will explore how SQL interacts with Big Data platforms, like Hadoop and Spark, and unlock the techniques to manage this voluminous data effectively. I'm not saying it will be a walk in the park, but armed with SQL and the insights from this book, you'll be well-equipped to take on the challenges and opportunities that Big Data presents.

L.D. Knowings

In the Ocean of Big Data, SQL Is Your Lifesaver

"Data is a precious thing and will last longer than the systems themselves."

— Tim Berners-Lee

The New Gold Rush: Swimming in the Ocean of Big Data

Picture the early days of the Internet, a vast, untamed wilderness of information. Fast-forward to today, and that wilderness has evolved into a sprawling, interconnected realm that we term Big Data. According to an IDC report, the collective sum of the world's data will grow from 33 zettabytes in 2018 to 175ZB by 2025, a compounded annual growth of 61%. Mind-boggling. And here's where the plot thickens: this colossal ocean of data is not just about volume; it's about the speed at which new data is generated and the myriad types of data we now deal with. No longer are we handling mere spreadsheets or databases; we're talking real-time analytics, social media metrics, sensor data, and so much more.

The Trusty Anchor in Big Data: SQL

So, how do we make sense of this labyrinthine world of Big Data? SQL. The same SQL you've used to query databases and generate reports is your go-to tool for mining this goldmine. While many might argue that Big Data requires specialized tools and languages, the fact remains that SQL continues to be an essential part of the Big Data ecosystem. You don't need to reinvent the wheel; you need to understand how to use SQL as your wheel in the complex machinery of Big Data.

The Swiss Army Knife of Big Data: SQL's Multifaceted Role

Remember when you thought SQL was just for data retrieval? Those days are long gone. SQL has grown up and joined the Big Data party in full swing. It's not just about SELECT queries anymore. Today, SQL can handle data streaming, unstructured data, and even machine learning algorithms, thanks to advancements in SQL-based platforms tailored for Big Data like Apache Hive and Google BigQuery. The beauty of SQL lies in its simplicity and its power. It's a language that's both easy to grasp and yet incredibly robust in its capabilities, which makes it a force to be reckoned with in the Big Data universe.

Practical Exercise: SQL Big Data Query Drill

Time for some hands-on fun. Grab a sample dataset from a Big Data platform of your choice. Try running an SQL query involving a complex join operation between two large tables. Not as daunting as you thought, right? You'll find that the principles you've learned in previous SQL exercises apply here, too. The only difference is the scale; as you'll discover, SQL scales beautifully.

Your Role in the Big Data Equation

So, where do you fit in all of this? Whether you're a data analyst in healthcare, a business consultant in finance, or a market researcher, the chances are that Big Data affects your work somehow. It offers both challenges and opportunities, but the common denominator is SQL. From cleaning messy data to integrating multiple data sources and even predictive modeling, your SQL skills are your compass in navigating this landscape.

The Road Ahead: SQL and Big Data in Harmony

So, what's next on the horizon? Integrating SQL and Big Data is an evolving field, with innovations cropping up frequently. There are now SQL functions tailor-made for Big Data analytics, and Big Data platforms increasingly offer SQL interfaces. The future is bright, and SQL is illuminating the way.

Thought-provoking Question: What Big Data challenges do you currently face or anticipate in your field? How could SQL serve as a solution to these challenges?

Chapter 8

Big Data: The Elephant and How SQL is its Tamer

"Data is not information, information is not knowledge, knowledge is not understanding, understanding is not wisdom."

— Clifford Stoll

Welcome to the Gigantic Zoo: Meet Big Data

You've probably heard about Big Data enough to feel its weight. But let's strip it down and understand what it is. The term Big Data encapsulates immense volumes of data that can be structured or unstructured. It's not just the data's size; it's also its complexity and the speed at which it is generated or processed. The industry often discusses Big Data in terms of the '5 Vs': Volume, Velocity, Variety, Veracity, and Value. It's a beast with many aspects, like an elephant with not just one but multiple trunks, each serving a different purpose. This creature roams in business, healthcare, science, and virtually any other domain. Imagine sifting through petabytes of medical records, customer interactions, or climate models. Overwhelming.

The Bumpy Ride: Challenges in Taming Big Data

Confronting this elephant, you'll find multiple challenges. Storage is often the first hurdle. Where do you keep this mountain of information? And then comes processing. It needs to be sorted, filtered, and analyzed to make data useful, which is easier said than done when dealing with Big Data. Then there's the issue of security. With significant data comes great responsibility. How do you ensure that the sensitive parts of your data are as invulnerable as Fort Knox? It's a dilemma of complexities. Yet, the compelling opportunities keep professionals like you and me intrigued.

The Golden Opportunities: The Silver Lining in the Big Data Cloud

Why would anyone willingly wade through this complex labyrinth? Because Big Data is a treasure trove. It provides customer insights for businesses that are nothing short of a gold mine. In healthcare, Big Data analytics can predict outbreaks, improve patient care, and even revolutionize drug discovery. In science, it has the power to solve some of the most intricate problems, from climate change to the mysteries of the universe. The opportunities are not just abundant; they are transformative.

SQL: The Elephant Whisperer

Now, this is where SQL steps in like a seasoned elephant whisperer. SQL stands out for its simplicity, versatility, and universal acceptance. When wrestling with Big Data, SQL is the guide that helps you sift through the data, make sense of it, and extract valuable nuggets of information. SQL is no longer just a tool for databases; it has grown to become a vital part of Big Data analytics. And don't forget, SQL's not just about querying data; it's also about updating it, manipulating it, and securing it. Whether you're a data analyst in finance or a researcher tackling climate change, SQL gives you the power to make data dance to your tunes—safely and efficiently.

Interactive Exercise: SQL Queries for Big Data

Let's get our hands dirty. Consider any Big Data platform you've worked with or want to explore. Create a complex SQL query that performs multiple operations—a JOIN, a GROUP BY, and maybe a window function. What do you notice about the performance? Surprisingly efficient, right? SQL doesn't just work; it works well, even when the data scales to Big Data proportions.

Why You Can't Afford to Ignore This

You're here because you don't just want to be good; you aim to be exceptional in your field. Knowing SQL in the world of Big Data isn't just a skill; it's a superpower. SQL provides the toolset to unlock the vast opportunities hidden in Big Data. It equips you to tackle challenges head-on, making you invaluable to your organization and setting you apart in your career.

SQL: The Unsung Hero in Taming Big Data

 "In God we trust. All others must bring data."

— W. Edwards Deming

The Invaluable Role of SQL in Managing and Analyzing Big Data

Ever stared at a sea of data and wondered, "How do I even start making sense of this?" Well, let me let you in on a secret weapon—SQL. SQL isn't just the bread and butter of relational databases; it's the Swiss Army knife for anyone dealing with Big Data. At its core, SQL gives you a syntax, a language, and a way to converse with your data. It allows you to ask questions in a format your data understands and can respond to. Think about it: if Big Data were an enormous jigsaw puzzle, SQL would be that friend who's ridiculously good at solving them. What are your frustrations with data storage, security, and processing? SQL has your back. It helps you parse terabytes and petabytes, narrow down what's relevant, and keep it safe from prying eyes.

Advanced SQL Techniques: Your Toolkit for Big Data

Now, SQL isn't just SELECT, FROM, and WHERE. It's far more nuanced, especially when you're handling Big Data.

Let's discuss the advanced techniques you'll want in your toolbelt.

Window Functions:

Imagine you're a healthcare analyst and have data on patient visits for the past five years. You need to find monthly visit trends and keep yearly metrics in sight. Here, SQL window functions come to the rescue. They let you perform calculations across table rows related to the current row within the same result set. You can find averages, run totals, or even ranking without altering your data. It's like having a lens to adjust to get the close-up and the big picture.

Partitioning:

Let's say you're working in finance and have transaction data flowing faster than a Wall Street ticker. SQL's partitioning feature lets you break this massive table into smaller, more manageable pieces called partitions. You can partition your table by range, list, or even hash. It's your organizational lifesaver, making data retrieval operations less overwhelming.

Indexing:

A database without indexes is utterly chaotic, like a library without a catalog. If you're handling Big Data, indexing is your non-negotiable. SQL allows you to create various indexes tailored to your specific querying needs, making data retrieval a breeze. Every millisecond counts in Big Data; indexes are your ticket to speed.

Hands-On Exercise: Advanced SQL Queries for Big Data

Alright, enough talking. Let's do a quick exercise to get these concepts ingrained. Open up any SQL editor you're comfortable with, and let's do some virtual lab work.

1. Window Functions: Write an SQL query that uses the RANK() window function to rank sales data for the last quarter.
2. Partitioning: Create a sample table of customer data and write a query to partition this data by region.
3. Indexing: Think of a table that could use an index and write the SQL command to create that index.

How did it go? Not as daunting as you thought, right?

Data Whisperers: SQL's Symbiotic Relationship with Data Mining and Machine Learning

 "Data is the new oil."

— Clive Humby

Data is the new oil; it fuels our decisions, powers our organizations, and drives innovation. But let's get real: merely having data isn't enough. Imagine owning an oil field without knowing how to extract the oil, refine it, or even where to start drilling. That's precisely how data is untapped and not fully understood in many databases. This is where data mining and machine learning come into play, acting like the expert drillers and refiners in our oil analogy. But what's the tool that makes these processes efficient? You guessed it—SQL.

The ABCs of Data Mining and Machine Learning

Before diving into how SQL fits this, let's clarify what data mining and machine learning are.

Data Mining

Data mining is the process of discovering patterns, correlations, and anomalies within large data sets to predict outcomes. Think of it as sifting through a haystack to find your needles; only in this case, the haystack is a database, and the needles are the insights you can turn into actionable strategies. It's like the detective work of the data world, where you're trying to find clues to solve business problems or create new opportunities.

Machine Learning

Machine learning, on the other hand, goes a step further. It's a type of artificial intelligence that enables systems to learn from data. So, instead of you sifting through the haystack repeatedly, you train a machine learning model to do it for you. Over time, the model gets better at finding those needles.

SQL: The Trusty Assistant in Data Mining and Machine Learning

SQL in Data Mining

SQL is indispensable when you're knee-deep in the data mining process. With SQL queries, you can quickly and efficiently analyze data to identify the patterns or trends you're interested in. For example, SQL provides specific commands like CLUSTERING and PREDICTION JOIN that can be extremely helpful in data mining tasks. These commands allow you to segment your data into clusters or predict future data points based on historical trends, all within the familiar SQL environment.

SQL in Machine Learning

SQL's role doesn't end at data mining. When it comes to machine learning, SQL can be used to create models, predict outcomes, and even evaluate the effectiveness of different models. Commands like CREATE MODEL and PREDICT are just some of the functionalities SQL has built into it for machine learning tasks. And the best part? You don't have to step out of your SQL comfort zone to use machine learning; it's all integrated.

Synergy with Other Tools: The Best of All Worlds

Now, SQL isn't a lone wolf. It knows when to take help and how to play well with others. Especially when you're tackling complex data mining and machine learning tasks, integrating SQL with tools like Python or R can be like adding rocket fuel to your data operations. Python and R have extensive libraries for advanced statistical analysis and machine learning. You create a powerhouse of data analysis capabilities by using SQL to manage and manipulate your data and then passing it through Python or R for more complex operations.

Your Toolbox: Practical Exercises

Alright, let's get those hands dirty, shall we? Fire up your SQL environment and try these tasks:

1. Data Mining: Use the SQL CLUSTERING command to segment a dataset of your choice into meaningful clusters.
2. Machine Learning: Use SQL's CREATE MODEL to create a simple machine learning model based on a dataset you're familiar with.
3. Integration: Try a simple Python script that pulls data from an SQL database, applies a machine learning algorithm from the scikit-learn library, and then pushes the results back into the SQL database.

How are you feeling? Empowered, I hope.

The Rubber Meets the Road: Hands-On SQL Techniques for Big Data

> *"Give me six hours to chop down a tree, and I will spend the first four sharpening the axe."*
>
> — Abraham Lincoln

Good ol' Abe Lincoln knew the value of preparation and skill honing, didn't he? Let's parallel that wisdom to what we're dealing with here—Big Data and SQL. By now, you understand the theories, the commands, and the strategies. But theories are just that—ideas—until you apply them. And this is where we roll up our sleeves, my friend.

The Exercise Laid Bare: What Are We Doing Here?

We're diving into an exercise designed to put your SQL knowledge to the test on a Big Data scale. We're taking those SQL commands, those techniques you've been studying, and applying them to an ocean of data. It's time to show that Big Data isn't too big for SQL or you.

Why this exercise? Because it's one thing to talk the talk and another to walk the walk. You've got to get your hands dirty with actual data, solve real problems, and see real results. After all, isn't that why you're here? To solve real-world issues in your job, make those advancements in your career, and impress your colleagues and bosses.

Chapter 8

A Step-by-Step Guide to Handling Big Data with SQL

Step 1: Get Your Data Source Ready

First, you'll need a dataset that qualifies as "Big Data." Kaggle has some excellent datasets you can download for free. For this exercise, let's choose a dataset that interests you and is relevant to your industry. Are you in healthcare? Maybe a dataset of medical records. Marketing? How about customer behavior data?

Step 2: Import the Data into Your SQL Environment

After downloading your dataset, the next step is to import it into your SQL environment. You can do this using your SQL software's IMPORT command or GUI interface.

Step 3: Explore the Data

Run some basic SQL queries to understand what your data looks like. Use the SELECT command to view different parts of the data and the COUNT function to see your dataset's size.

Step 4: Clean the Data

Here's where the grunt work comes in. You know what I'm talking about: those pesky null values, the duplicates, and the inconsistencies. Use the UPDATE and DELETE commands to clean up the data. You'll thank yourself later.

Step 5: Run Your Queries

This is the fun part. Start running some complex SQL queries to analyze the data. Use the techniques discussed in previous chapters, like JOIN operations, GROUP BY clauses, and aggregate functions. The world—or, in this case, the dataset—is your oyster.

Step 6: Reflect on the Output

Take a look at the output of your queries. Do they make sense? Are you gaining the insights you were hoping for? If not, it might be time to return to the drawing board and tweak your queries.

Handy Resources to Elevate Your Skills

If you're looking to expand your horizons even further, there are additional resources that can deepen your understanding of SQL's capabilities with Big Data:

1. Computer PDF: Offers exercises deeper into Big Data analytics.
2. Tech Target Blog: Provides insightful articles that debunk myths surrounding SQL's capacity to handle Big Data.

Chapter 8

Your Toolkit Isn't Complete Without Real-World SQL Tactics

 "Tools are great, but the craftsman counts."

— Andy Grove, Co-founder of Intel

Ah, Andy Grove, a man who knew a thing or two about crafting intricate systems, isn't far off the mark. You see, learning SQL isn't just about acquiring a new skill; it's about becoming a craftsman in a world that increasingly values data-driven decisions. Now that you've amassed a SQL commands and techniques toolset, what's next? Ah, the natural world awaits, my friend. You've got an arsenal of SQL skills; now it's time to take them out for a spin in the grand arena where theory meets practice.

Real-World SQL: Bridging the Gap

Let's get candid for a moment. You've probably heard or read about SQL enough to know its merits, but if you're like most people, you're secretly asking, "When am I going to use this in my day job?" Sound about right? Trust me, I get it. You're a data analyst, a business researcher, or perhaps even a healthcare professional who's been told that SQL is the Holy Grail. You've got responsibilities, deadlines, and targets to meet, and you want to know how SQL fits into this puzzle.

Well, this is your pit stop. Think of it as that moment when a Formula 1 racer pulls over to change tires. You've been on this track, going at top speed, learning SQL commands, and understanding data types, and now, you're stopping for a moment to fit those winter tires for the slippery road of real-world applications ahead.

What's Coming Up: A Sneak Peek

Suppose you're wondering whether SQL can help you make sense of that massive, convoluted dataset you're dealing with at work or how it can help you make more informed project decisions. In that case, the next chapter is your playground.

You can anticipate a deep dive into real-world applications of SQL across various industries. Whether you're crunching numbers in finance, making marketing strategies, or analyzing patient data in healthcare, there's something tailor-made for you. We're talking about case studies that aren't just hypothetical exercises but are drawn from real-world challenges that professionals like you face daily.

How do you ensure the quality of data across diverse data sources? How do you integrate disparate sets of data into a unified whole for analysis? How do you navigate the labyrinth of data exploration and visualization? All these questions and more will find their answers.

Chapter 9
When SQL Meets the Real World: A Symphony of Applications Across Industries

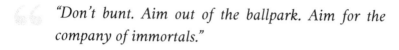 *"Don't bunt. Aim out of the ballpark. Aim for the company of immortals."*

— David Ogilvy, Father of Advertising

You've honed your SQL skills, deciphered complex queries, and even dabbled in advanced analytics. It's been a rewarding experience so far. But let's cut to the chase; you're probably wondering how all this SQL wizardry applies to your industry, projects, and, most importantly, your career.

Practical Applications in Various Industries

Let's spill the beans—SQL is not just another programming language that computer geeks ramble on about. It's a universal language spoken in the hushed halls of healthcare, the bustling floors of financial markets, the trendy offices of

L.D. Knowings

marketing agencies, and the secure command centers of cybersecurity. SQL is your ticket to any job and the position that makes a difference.

Finance: A Tale of Billions and SQL

Let's talk money. SQL is as crucial as the dollar signs on a balance sheet in the finance sector. Imagine handling transactions that run into billions, a slight error, and you're talking about an explosive cascade of problems. SQL allows for the seamless management of these transactions, ensuring accuracy down to the last cent. But it's not just about transactions but risk assessment, fraud detection, and portfolio management. SQL here is like your most reliable financial advisor but at the speed of light.

Healthcare: Lives in the Balance

Now, shift gears and think about healthcare. In a hospital, every second counts. Imagine the convenience of accessing patient records, medical histories, and treatment plans with a few clicks. SQL makes that happen. It enables healthcare providers to manage vast amounts of patient data efficiently, ensuring that doctors and nurses have the information they need precisely when needed. The impact? Quicker diagnoses, more effective treatments, and lives saved.

Retail and Marketing: The Customer is King

If you're in retail or marketing, you already know that understanding your customer is half the battle won. SQL helps you do just that. From managing inventories to tracking customer behavior, SQL plays a pivotal role. Imagine predicting what products will fly off the shelves next month. With SQL, you're managing a business and foreseeing its future.

Technology and Cybersecurity: The Shield and Sword

SQL is both a shield and a sword in the realm of technology. On the one hand, it helps manage complex databases that power your favorite apps. On the other, it's a vital tool in cybersecurity, helping identify vulnerabilities and protect against data breaches. You might not see SQL in action here, but rest assured, it's working round the clock to keep things secure.

Impact and Benefits Across the Board

So, why should you care about SQL's versatility? Because its applications are as varied as they are impactful. In finance, it minimizes risk; in healthcare, it saves lives; in marketing, it maximizes ROI; and in technology, it safeguards data. These aren't just tasks; these are transformations—turning raw data into actionable insights and insights into impactful decisions.

Exercises and Self-Assessments

Before we move on, here's a quick exercise. Take a few minutes to jot down some of the tasks you handle in your job role that involve data. Next, think about how SQL could make those tasks more efficient. This exercise isn't just about understanding SQL's applicability; it's about envisioning your growth and efficiency in your career.

Harnessing SQL's Might: Real-World Problem Solving Unveiled

"The value of an idea lies in the using of it."

— Thomas Edison

You've been neck-deep in SQL commands, probably lost in a maze of SELECT statements and JOIN clauses. You must now wonder how these abstract concepts translate into real-world applications that make a difference. Let's pull back the curtain and show you SQL's true prowess in addressing problems that matter in industries and scenarios where data isn't just numbers but the lifeblood of decision-making.

Real-world Problem Solving with SQL

Fraud Detection: SQL as a Watchdog

Here's something you may not have considered: SQL can be a formidable tool in preventing fraud. Financial institutions are particularly vulnerable to fraudulent activities. Think about credit card fraud or identity theft; we're talking about millions, if not billions, at stake. SQL has a way of sifting through complex data to identify irregular patterns or anomalies that may signify fraudulent activity. You can flag suspicious transactions in real time using SQL functions such as clustering and classification. Imagine the power and responsibility that comes with that. You're not just writing queries; you're writing the code that keeps people's hard-earned money safe.

Predicting Customer Trends: SQL as a Soothsayer

In marketing and retail, predicting customer behavior is the Holy Grail. It's one thing to know what your customer bought last week; it's another to indicate what they'll want next month. SQL allows you to analyze historical data, look at purchasing patterns, and predict future trends. With SQL functions like time-series analysis, you can project customer behavior into the future. You're no longer reacting to the market but ahead of it. Can you feel the thrill of that power?

Improving Operational Efficiency: SQL as the Orchestrator

Let's focus on operational efficiency, the unsung hero of business success. SQL is pivotal in streamlining operations, whether inventory management, workforce optimization, or supply chain logistics. Complex SQL queries can analyze multiple variables to recommend the most efficient routes for delivery trucks or the optimal inventory levels to maintain for each product. What this means for you is a significant cost reduction and a spike in productivity.

Practical Exercise: Pinpointing Real-world Problems

To help you apply this knowledge, consider this exercise. List some real-world challenges in your industry or job role involving data. Now, think of the SQL commands or techniques discussed in previous chapters that could solve these problems. This exercise isn't just academic; it's about envisioning SQL's direct impact on your professional life.

SQL in the Trenches: Where the Rubber Meets the Road

"You can have data without information, but you cannot have information without data."

— Daniel Keys Moran

Ah, SQL. For some, it's just a tool, a means to an end. But for you and me, it's more than that—the key unlocks the treasure chest of data, turning raw numbers into actionable insights. We've reviewed the basics, dissected the queries, and played around with functions. But now, let's pivot. We will dive into the real world, where SQL isn't just code—it's a problem solver.

Chapter 9

Effective Use of SQL in Different Scenarios

Edgenet: Real-Time Retail Product Data

First up, Edgenet is a company with a problem many retail and e-commerce platforms face: how to gain real-time access to an ever-changing array of retail product data. They turned to SQL with in-memory technology to tackle this challenge. The SQL solution was a game-changer, employing lightning-fast data retrieval methods that provided real-time insights into product availability, pricing, and customer preferences. The outcome? A dramatically streamlined operation that could adapt to market trends in real-time, giving Edgenet a significant competitive edge.

OpenText: Enterprise Information Management

Let's shift gears to OpenText, a global leader in Enterprise Information Management (EIM). They were grappling with the complexity of managing vast amounts of unstructured data. SQL came to the rescue with its robust data transformation and integration capabilities. The SQL implementation enabled the unstructured data to be managed and allowed for effective data analytics, leading to more intelligent business decisions for OpenText.

Meta: Static Analysis of SQL Queries

Now consider Meta—yes, the social media behemoth. Meta engineers were faced with a unique problem: how to conduct static analysis of millions of SQL queries to ensure optimal performance. The SQL solution they implemented was an innovative mix of query optimization techniques that significantly improved the performance and reliability of data operations. This was no small feat; we're discussing enhancing the user experience for billions of people worldwide.

SAP on SQL Servers: Enterprise Resource Planning

Next, let's look at how SQL is shaking up the world of Enterprise Resource Planning (ERP) with SAP on SQL servers. SQL has been instrumental in migrating SAP solutions to the cloud, optimizing resource allocation, and improving overall efficiency. Companies using this SQL-based approach have experienced operational improvements and substantial cost savings—a double win in any book.

DocuSign: E-Signature Verification

DocuSign, the electronic signature company, had to verify the authenticity of thousands of e-signatures daily. The company employed SQL with SSIS (SQL Server Integration Services) to automate the verification process. This SQL solution drastically reduced time and human resource investment, making the verification process efficient and secure.

RHI Magnesita: Manufacturing and Supply Chain

In the manufacturing sector, RHI Magnesita faced the challenge of optimizing its global supply chain. Using SQL-based solutions, they could integrate real-time data from multiple sources, resulting in a more responsive and efficient supply chain. The improvements in logistics and inventory management were immediate and substantial, affecting the bottom line positively.

Health-Data Analysis: Medical Research

Lastly, look at health data analysis, a sector where the correct information can save lives. Researchers have been using SQL to sift through enormous datasets to find patterns that could lead to medical breakthroughs. Imagine using SQL to help identify the most effective treatment for a particular disease or pinpoint an outbreak's cause. That's the kind of real-world impact we're talking about.

The SQL of Tomorrow: The Blueprint of Your Success

"To be successful tomorrow, you have to have a good foundation today."

— Unknown

Allow me a moment to dust off my track coach hat. In athletics, it's one thing to understand the mechanics of a race; it's another thing altogether to cross the finish line first. This chapter is for those who aren't content just knowing the SQL commands. You want to know how to turn this skill into a fulfilling and profitable career. So, let's discuss how you can make SQL your profession, not just your passion.

Tips to Building a Career in SQL

Job Roles: The Many Hats You Can Wear

First things first: understanding the range of job roles that require SQL expertise. It's not just about being a Database Administrator anymore. The opportunities are endless for data analysts, business intelligence analysts, and data scientists. Think of these roles as the different races in a track meet. In the 100m dash, pure speed wins, but in the 800m, it's a blend of speed and endurance. Similarly, while a Data Analyst might focus on quick data retrieval and primary analysis, a Business Intelligence Analyst will delve into deeper layers, decoding data for strategic decisions. And let's not forget Data Scientists, the decathletes of the data world, skilled in SQL and many other data manipulation and machine learning tools.

From Novice to Ninja: The Growth Chart

Starting your career in SQL isn't just about landing that first job. It's about growth. You don't want to be that guy who can only write basic SELECT queries for the rest of your life. Remember, the tech industry evolves faster than a sprinter out of the blocks. You have to keep up. Don't just settle for the basics; make it a habit to learn something new every day. Join online forums, follow industry experts, and engage in SQL challenges. Keep your skills honed and updated.

How to Get Real-World Experience

Let's face it: Employers don't just want to know you can write SQL queries; they want to see you've applied these skills in real-world situations. This is the same in almost any profession—theory is vital, but application is king. So, how do you gain experience before landing your first job? Open-source projects are a great way to get your feet wet. These projects allow you to work on actual tasks and offer something tangible to show prospective employers. Another avenue is internships. Companies often look for interns who can handle data manipulation tasks. It might not pay much (or at all), but the experience is priceless.

Certifications and Continuous Learning

You've probably seen those LinkedIn profiles with a laundry list of certifications. While you don't need to go overboard, having a few reputable SQL certifications can provide an edge in a competitive job market. But remember, certificates are not a one-and-done thing. The tech world doesn't stand still, and neither should your learning. Platforms like DataCamp and Simplilearn offer a range of courses that can help you go from SQL novice to expert.

Networking: Your Secret Weapon

Yes, even in the tech world, who you know can be just as important as what you know. Networking doesn't mean schmoozing at boring parties but connecting with people who share your interests and can offer valuable insights or opportunities. Attend industry conferences, join SQL user groups, and don't underestimate the power of social media. A simple tweet or LinkedIn post can sometimes open doors you didn't even know existed.

The Summit of SQL: Where You Stand Now and Beyond

"The best way to predict the future is to create it."

— Peter Drucker

As we approach the end of this transformative experience, let's pause for a moment. Breathe in. Reflect. You've amassed a staggering range of SQL skills, from basic commands to advanced data analytics techniques. You're no longer the person staring nervously at a database, wondering how to make sense of its arcane symbols. Now, you're sculpting these symbols into meaningful narratives, deciphering the hidden language of data to inform and elevate decision-making.

Chapter 9

A Panorama of Your Progress

You've conquered the steep slopes of data cleaning and preparation, a critical task often occupying more time than analysis. Remember the frustration of dealing with messy, inconsistent data? Now, you've got the SQL tools and techniques to clean it up and shape it into something valuable. You've also navigated the intricate pathways of data integration, merging diverse data sources into a coherent, unified dataset. That's not just a technical skill; that's a form of art.

The Real-World Canvas for Your SQL Skills

The unique blend of theoretical and hands-on training you've received from this book is your armor, toolkit, and palette of colors as you paint your career in vibrant hues. It's not about SQL commands in isolation; it's about how to wield them to solve real-world problems, from simple queries to complex data visualizations. What you've learned here directly applies to your job, whether you're a healthcare data analyst, a finance business researcher, or a machine learning engineer in technology. Your newfound SQL prowess is versatile and applicable across multiple industries and roles.

Lifelong Rewards and Continual Growth

Let's face it: the tech field is ever-evolving. What's trending today may be obsolete tomorrow. But SQL? The timeless classic, the evergreen skill, continues to be the backbone of data manipulation and analysis. The learning doesn't stop here; it can't. Technology will continue to advance, and with it, the capabilities of SQL will expand. Keep up-to-date with industry trends, participate in online communities, and attend workshops and seminars. The SQL realm is vast and ever-expanding, and you have the key to unlock its endless potential.

The Gift That Keeps on Giving

Mastering SQL isn't just about you. It's about how you can contribute to your organization, solve complex problems, and make impactful decisions. It's also about how you can mentor others, sharing your SQL knowledge and skills with colleagues or perhaps the next generation of data enthusiasts. You see, mastering SQL is a gift that keeps on giving. It enriches your professional life and contributes positively to those around you and the organizations you'll serve.

The Power You Hold: Realizing the Impact of Your SQL Skills

 "Do not wait to strike till the iron is hot, but make it hot by striking."

— William Butler Yeats

Allow me to pause and applaud you for a moment. You've ventured deep into the labyrinth of SQL and data analytics, and you're emerging with the spoils of your efforts: a comprehensive skill set that's both actionable and robust. You've pushed through the initial barriers, tackled the common issues that plague many in the data analytics field, and now stand on a vantage point that offers you a broad view of possibilities.

Your Toolkit: More Than Just Commands

Remember the hours you spent mastering data cleaning? It's now an investment yielding dividends. You're not just removing null values or handling missing data; you're crafting the data quality any analyst would envy. And let's not forget about data integration. You've moved beyond the phase of being intimidated by different data formats and sources. You're the maestro, orchestrating these disparate parts into a harmonious dataset.

The Canvas of Opportunities: Where to Apply Your Skills

I can't stress enough how these skills are not just theoretical achievements to be gloated over in a resume. No, they're your ticket into many industries—healthcare, finance, technology, you name it. Imagine applying your capabilities to decipher the human genome or to predict stock market trends. These aren't mere SQL commands; they're tools of transformation.

The Value of Continuous Learning in a Dynamic World

The world of tech evolves at a dizzying speed. While the core of SQL has remained robust over the years, the periphery keeps expanding. So, don't let the ink dry out; keep writing new chapters in your SQL story. Stay curious. The best analysts are those who view learning as a perpetual process. The industry won't wait, so why should you?

Be the Catalyst in Your Organization and Beyond

Your skills have an exponential impact. By mastering SQL, you're enhancing your career and contributing to your organization's decision-making prowess. But the ripple effect doesn't stop there. You're now in a position to mentor to spread the knowledge. Think of it as compound interest; the more you invest in others, the more the community grows, and the returns are manifold.

Chapter 9

"The Future Belongs to Those Who Prepare for It Today"

— Malcolm X

Alright, let's take a breather. You've been wading through SQL statements, wrestling with data structures, and shedding old misconceptions about the "techiness" of data analytics. So, how do you feel? Empowered? Overwhelmed? Let's sift through these emotions, shall we?

The Alchemy of SQL: Transforming Data into Wisdom

You started with the basics—those fundamental SQL commands and syntax. But then, something almost magical happened. The focus shifted from mere commands to molding raw data into actionable insights. It's as if you've been handed an ancient scroll detailing the art of alchemy; instead of turning lead into gold, you're converting data into wisdom.

The Arena of Applications: It's Not Just About Tech

Remember, SQL isn't some arcane language meant only for Silicon Valley wizards. It's the Swiss Army knife of the data world. Whether you're in healthcare, finance, or marketing, these skills are universally applicable. The ability to make data-driven decisions is like having a secret weapon in your professional arsenal.

Here and Now: Keeping Pace with Ever-changing Tech

Ah, the only constant in tech is change. New libraries, updated tools, different methodologies—the landscape is ever-evolving. The good news? You're well-prepared. You have the foundational skills; it's about refining and expanding them. Stay thirsty for knowledge; the elixir keeps you relevant in a field that never sleeps.

Don't Just Be a Player; Be a Game-Changer

So, you've honed these skills to enhance your career. That's brilliant! But have you considered the ripple effect? The decisions you influence could lead to groundbreaking changes in your organization. Moreover, you're now in a position to mentor others, to pass on the baton. It's a cycle of growth where everyone benefits from your expertise.

The Elephant in the Room: Confronting Our Hidden Desires

We all have those hush-hush aspirations we dare not speak about, even in our most private moments. Maybe it's solving an industry-wide issue or pioneering a new data analytics method. With the skills you've gained, these aren't far-fetched dreams; they're goals waiting to be achieved. So, why not aim high?

Chapter 9

The Naysayers and the Trailblazers

Let's address those annoying myths that say mastering SQL is akin to climbing Mount Everest without gear. Not true. You've seen firsthand that the subject is complex but conquerable. You don't need a Ph.D. in mathematics or a background in programming to excel here. What you need is the willingness to learn and the persistence to apply.

To Infinity and Beyond: The Horizon of Possibilities

You're not just collecting skills; you're gathering tools for a lifetime. The SQL queries you draft today could be the foundation for a future project revolutionizing your industry. As you've seen, SQL isn't just a tool; it's a mindset that equips you to tackle any data challenge thrown your way.

The Evergreen Value of Mastery: A Lifelong Asset

By now, you should be aware that mastering SQL isn't a short-term sprint; it's more of an ultra-marathon. The finish line keeps moving, but every step you take enriches you professionally and personally. Your enhanced problem-solving skills, ability to generate insights, and newfound proficiency in data analytics are not mere resume boosters —they're life enhancers.

Afterword

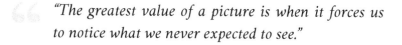 *"The greatest value of a picture is when it forces us to notice what we never expected to see."*

— John Tukey, American mathematician and
the founder of exploratory data analysis

Breaking Down Barriers

By now, you've developed a deep, intricate relationship with SQL and data analytics. You've moved past the myths and misconceptions that once held you back. The notion that mastering SQL and data analytics requires a Ph.D. in rocket science? We've obliterated that fallacy. Together, we've shown that anyone—whether you're a data analyst, business strategist, or a healthcare professional—can unlock the astonishing powers of data.

The Heart of the Matter

The essence of this book boils down to one vital reality: SQL and data analytics are indispensable tools for the modern professional. These are not just lines of code or strings of numbers; they are the lifeblood of intelligent, informed decision-making. You've learned not just the syntax but the applicability and relevance of SQL in today's digital age. You now have the keys to unlock patterns and insights that were once invisible, buried deep within spreadsheets and databases.

What You Take Home

You've grasped the techniques, the approaches, and the mindset needed to excel in data analytics. This book gave you more than just commands; it offered you a whole new lens through which to view your world professionally and personally. You've gained the skills to clean your data, integrate various data sources, explore and visualize complex datasets, and apply statistical techniques. The immediate outcome? A more valuable thank you. What is the long-term effect? A transformation not just in how you work but in how you think.

Looking Forward

Trends in technology, artificial intelligence, and machine learning are not slowing down. The world is becoming more data-centric by the minute. But with the knowledge you've acquired, there's no reason to be left behind. You're now equipped to navigate the complexities of any data-driven challenge that comes your way.

Breaking the Mold

This is not just another book on SQL or data analytics. This is your new toolkit, a new set of gears, and a roadmap to professional excellence. It's your unique advantage in a sea of competitors. You didn't just read a book; you transformed. You've transcended many hurdles, from data integration to quality assurance. You are now part of an elite group of professionals who understand data and can wield it to craft narratives, solve problems, and drive impact.

A Toast to Your Triumphs

There's nothing quite like the euphoria that washes over you when a piece of SQL code runs flawlessly, especially after hours of scratching your head, rereading lines, and double-checking syntax. It's akin to hitting a home run or sinking a three-pointer at the buzzer. It's the essence of the "Ah-ha!" moment, where everything clicks into place, and you suddenly feel invincible. That, my friend, is not just the joy of problem-solving; it's the thrill of mastery. It's not only about understanding SQL or data analytics; it's about knowing that you've conquered a skill that can shape industries, drive innovation, and create opportunities.

Afterword

Your Success Story is Just Beginning

Think about Sarah, a business analyst in the healthcare sector, who used the SQL skills she acquired to optimize hospital resource allocation significantly. Armed with the ability to make data-driven decisions, she became an unsung hero during a challenging period, ensuring that staffing and supplies were where they needed to be, precisely when they needed to be there. Sarah's story isn't an isolated incident; it's a testament to the transformative power of SQL and data analytics. The skills you've acquired have a real-world impact beyond rows and tables, screens, and keyboards.

The Power of a Single Line of Code

You've learned the nitty-gritty details, the methods, the reasoning. You've battled messy data, tangled with inconsistent formats, and emerged victorious. A line of SQL code isn't just a command; it's a magic wand in your hands, capable of revealing insights that can lead to ground-breaking changes. The skills you now possess are a treasure for you and the communities and industries you'll impact.

So, what's next? A simple but impactful request: Use it. Use the knowledge you've gained, the skills you've honed, and the techniques you've mastered. Transform data from mere numbers into narratives, from statistics into stories. Whether in healthcare, finance, marketing, or any other field that relies on data, the world needs your skills. Translate the complex into the simple, the abstract into the tangible, and the theoretical into the practical.

Double Down on Your Investment

You invested your time and energy into this book and, by extension, into yourself. Now's the time to double down on that investment. The stage is set, the tools are in your hands, and the spotlight is on you. SQL is your newfound super-power; with it, you're bound to succeed and destined to excel.

One More Thing

If you found this book valuable, enlightening, or even life-changing, **consider sharing your thoughts in a review**. Your feedback helps us and guides others in choosing resources to help them transform their professional lives.

Now, go on. Use those SQL skills to uncover the hidden secrets in data and keep pushing the boundaries of what you thought was possible. Your success story is waiting to be written.